Kudos

"Jack Ewing is one of the best storytellers that I know and he has led an extraordinary life. For that reason alone he is well worth listening to. In his books, Jack established himself as a master of the short essay genre, telling true stories from his life that simultaneously entertained and imparted principles of tropical conservation biology. He communicates ecological concepts in such an easy style that the reader doesn't even realize they are being educated. For many years now I have made his first book, *Monkeys Are Made of Chocolate,* required reading for my tropical ecology students. Once they start reading, they can't put it down and it quickly becomes their favorite text. Jack's new book, *Monkeys Are Made of Mangos,* picks up where his previous book left off, with more delightful short stories that highlight the beauty and intricacy of the natural world as well as the twists and turns of his own fascinating life in Costa Rica."

Mike Mooring, Professor of Biology, Point Loma Nazarene University

"Imagine walking along the path from your house surrounded by the peace and beauty of the Costa Rican rainforest. A lovely afternoon, filled with bird song and the sound of the surf whispering through the trees. As you approach the remnants of the teak plantation, the opening in the forest reveals what you have come for: a lush, beautiful mango tree, teaming with fruit. You walked here so you could bring home two or three fresh mangos for lunch. Suddenly, the peace of the jungle is no more. It is broken by the animated screeching of white-faced capuchin monkeys as they throw mangos at you. Surprised, you retrieve your mangos and fade back into the rainforest, where you pause to chuckle a little, shake your head and exclaim, 'And we thought Monkeys are Made of Chocolate, it looks like they are also made of Mangos!' This monkey encounter is the inspiration for the first story in Jack Ewing's delightful new book, *Monkeys Are Made of Mangos*, written in the tradition of his first book, *Monkeys Are Made of Chocolate*. Once more we are invited by Jack to share his magical world as we follow him through the Costa Rican rainforest. Who knows what we may discover in the next clearing."

Pamela Herring, University of Texas at Brownsville

Also by Jack Ewing

Monkeys Are Made of Chocolate

"Jack Ewing's thirty-year adventure in a Costa Rican jungle has produced a book full of infectious love and amazing lore."

Daniel Quinn, award-winning author of *Ishmael*

"I taught about leaf cutter ants to my sixth grade class and the children were thrilled and amazed. Your pieces on Baird's tapir were so good I simply read them to the class, out of which they decided to raise money to adopt a tapir. Thank you!"

Stu Summer, middle school teacher - Hillsdale, New York

"Written in language accessible to everyone, the 32 stories in this book delve deep into the fascinating world of Costa Rica's tropical wildlife, skillfully intertwining ecological facts with current environmental and social issues which affect all of us. Even trained scientists will find something new in the personal, and often humorous, observations in these pages. This book is perfect reading for every inquisitive traveler to Costa Rica."

Rob Rachowiecki, *Lonely Planet Costa Rica* guidebook

"A fascinating collection of stories and essays from Jack Ewing's decades of observation of tropical flora and fauna. This book is the fun way to learn about biological corridors and the interrelatedness of all creatures."

Beatrice Blake, author of *The New Key to Costa Rica, 17th edition*

"*Monkeys are Made of Chocolate* is a tapestry of stories as rich as the land from which they came. Whether you like reading about huge snakes which always seem to come in pairs, the intelligent behavior of sloths, the ancient craft of boat building by digging out a tree trunk, or how toucans aren't quite as cute as they appear, this book is a treasure trove of Costa Rican life and natural history."

Georgie Wingfield, Agronomist, England

Also by Jack Ewing

Where Tapirs and Jaguars Once Roamed

"For Jack's many fans, the stories recalling episodes of his 45-year, personal evolution from cattle rancher, to emerging naturalist and conservationist, to renowned environmentalist will be equally enlightening and entertaining. This book will appeal to all ages and a wide range of readers… tourists, naturalists, environmentalists, natural history students, local history buffs and anyone who has had the pleasure of experiencing the natural Costa Rica that Jack has worked so hard to foster and protect.

Dorothy MacKinnon, travel writer – *Fodor's Guide to Costa Rica, Insight Guide to Costa Rica, Tico Times*

"Because of Jack Ewing and people like him we are sure to see the return of the tapir and jaguar to this extraordinary biological corridor, keeping this part of Costa Rica truly the 'rich coast' as it was meant to be. *Where the Tapirs and Jaguars Once Roamed* is a must read for those that love Costa Rica and those yet to discover this beautiful country."

Pat Cheek, Quepolandia

"The Costa Rican conservation story is a fascinating one and Jack Ewing has been an incredibly effective participant in its success. The book gives a very real and personal context to it. And he's a great story teller as well. I loved it!"

Richard Andrus, Professor, Binghamton University, New York

"*Where Tapirs and Jaguars Once Roamed* transports the reader into Costa Rica with an intimacy born from the author's life there and with a vividness that would be difficult to match even seeing it for yourself. It is fascinating and personal, never dry or academic, yet by the end there's no denying it's a thoroughly educational read."

Ray & Sue Krueger-Koplin, Toucan Maps Inc.

Monkeys Are Made of Mangos

Colorful, Captivating Costa Rica

JACK EWING

PixyJack Press, Inc.

Monkeys Are Made of Mangos: Colorful, Captivating Costa Rica

Published by PixyJack Press, Inc.
PO Box 149, Masonville, CO 80541 USA
www.PixyJackPress.com

print ISBN 978-1-936555-67-3
Kindle ISBN 978-936555-678-0
epub ISBN 978-936555-69-7

Library of Congress Cataloging-in-Publication Data

Names: Ewing, Jack, 1943- author.
Title: Monkeys are made of mangos : colorful, captivating Costa Rica / Jack Ewing.
Description: Masonville, CO : PixyJack Press, Inc., [2024] | Summary: "Insights and observations into the interconnectedness of life in Costa Rica, a place of constant change, where for over 50 years, author Jack Ewing has immersed himself in the ecology and restoration of this tropical environment"-- Provided by publisher.
Identifiers: LCCN 2023057882 (print) | LCCN 2023057883 (ebook) | ISBN 9781936555673 (trade paperback) | ISBN 9781936555680 (kindle edition) | ISBN 9781936555697 (epub)
Subjects: LCSH: Animals--Costa Rica. | Rain forest ecology--Costa Rica. | Rain forest conservation--Costa Rica. | Natural history--Costa Rica. | Ewing, Jack, 1943-
Classification: LCC QH108.C6 E953 2024 (print) | LCC QH108.C6 (ebook) | DDC 333.75097286--dc23/eng/20240129
LC record available at https://lccn.loc.gov/2023057882
LC ebook record available at https://lccn.loc.gov/2023057883

Cover illustration by Jan Betts.
Green ibis (back cover) by Jack Ewing.
Book design by LaVonne Ewing.

To my late friend,
Walter Odio Victory

Walter and I knew each other for 50 years. In the 1960s we hadn't met yet, but we were both earning our bachelor's degrees at universities in the United States, he at LSU and I at CSU. In the 1970s we were both stewards of large cattle ranches, Rancho La Merced and Hacienda Barú, each of which contained an abundance of biodiversity, beaches, rivers, mangrove estuaries, pastures, wetlands, and rainforests. In the 1980s we both became environmentalists and rewilders, as well as founding members of an organization that grew, matured, and later became known as Asociación Amigos de la Naturaleza del Pacifico Central y Sur (ASANA). In the 1990s we became partners in the ecological tourism agency SelvaMar.

Walter and I were both on the board of directors of ASANA in the year 1990 when we voted to create a wildlife corridor that would extend from the Savegre River to the Térraba River, thus connecting the Osa Peninsula with the Los Santos Forest Reserve.

Shortly after that Walter and some friends did a grueling 25 km hike beginning at Santa María de Dota, crossing the rugged and pristine Los Santos Forest Reserve, and ending at the village of Brujo. During the hike they saw tracks of Baird's Tapir, the largest land mammal in Central America. Upon returning home Walter contacted me by radio. "I have a name for the wildlife corridor," he blurted out. "Let's call it 'The Path of the Tapir'."

And so we did.

Map of Costa Rica

CONTENTS

PREFACE

When I came to Hacienda Barú in 1972, it was a cattle ranch with approximately half of the land in pasture or farmland and the other half in natural vegetation. The word "rewild" hadn't yet been coined in the late 1970s when we started returning land to Mother Nature and letting Her bring back the wild species.

Today, on Hacienda Barú's 330 hectares (815 acres), we have sighted and identified 369 species of birds. That is about 45 percent as many bird species as there are in the USA and Canada combined. We have 85 species of mammals, 96 species of amphibians and reptiles, 194 species of trees, 47 species of orchids, and over 10,000 species of insects.

In 1987, I started a list of all the birds sighted on Hacienda Barú nature reserve. It grew steadily at first, several months for the first 100 species, and several years for the next 100. Some of those are new to Hacienda Barú: crested guan, rufous-necked wren, scarlet macaw, Montezuma oropendola, and probably a few others that we assumed had always been here, but we just hadn't seen them and put them on the list.

The first spider monkey was sighted in 1997. It had been 50 years since they all died, along with the howler monkeys, during an epidemic of yellow fever. After the arrival of the first spider monkeys, we started a mammal list. In addition to the species we already knew about, squirrel monkeys, pumas, howler monkeys, and numerous species of bats have been added to the mammal list.

In October 2012, I spotted bird #361: a large, dark-colored bird with a curved bill sitting on the branch of a cecropia tree. I barely had time to dash into the house, grab my camera, run back out, and snap two photos before it flew away. The photos wouldn't have won any awards, but they were good enough to identify the bird as a green ibis (*Mesembrinibis cayennensis*). The species wasn't seen again until November 2022, ten years later. Two of them were walking around in the rain in an open area behind our house, plunging their long, curved beaks into the soggy earth all the way up to the feather line, and pulling back just in time to avoid getting dirt in their eyes.

As the sun was peaking over the coastal ridge the following morning we heard a new bird sound, a very pleasing one, and after a bit of searching spotted a single green ibis in a tree near the house. Stiles and Skutch in *A Guide to the Birds of Costa Rica* describe the call perfectly as "deep, mellow, rolling trills."

"That looks like the male," I commented to Diane. "I hope he has romance on his mind." Though both sexes have almost identical coloring, the male is fuller and heavier, while the female is longer and trimmer. The mornings had been rainy the first few days the ibis were here, and I had to use my imagination to see a slight tinge of green. I even mistook the male for a black vulture one cloudy day when it flew over my head and landed not far away. A brief opening appeared in the clouds; the ibis spread its wings to the warming rays of the sun, and finally I saw clearly iridescent green reflecting off its feathers and accepted that the bird was aptly named.

On the first dry day since the appearance of the green ibis we saw one of the birds carrying several thin sticks toward a thicket of trees, and the next day observed three more flights with nesting material. We never found the nest, but one morning I heard the call and located both birds in their favorite tree. They were standing close to each other and clashing their beaks together. He stopped, arched his neck, ruffled the feathers on the back of his head, and looked down on the female who was bowing low. Part of the mating ritual of the green ibis? Or just a lover's spat?

We never found the nest and don't know if they raised a chick. We do know that the following year there were more green ibis than the previous year, and neighbors up and down the coast were reporting seeing them. I believe we have a new, very beautiful resident bird species on Hacienda Barú and the surrounding area.

This is but a brief description of Mother Nature's handiwork at Hacienda Barú where many of the stories in this book took place. It is one of the few places in the world where biodiversity is steadily but surely increasing. That I have been able assist in some small way in this monumental endeavor is rewarding beyond words. But words are what we use to convey concepts and observations, and also feelings, and it is my hope that the words in this book are sufficient to arouse within you the sense of wonder that we at Hacienda Barú experience on a daily basis.

— Jack Ewing
November 2023

1. Monkeys Are Made of Mangos

And Lots of Other Stuff

Hacienda Barú is much different today than when we first moved here in the early 1970s. Living with the rewilding of the land from cattle ranch to nature reserve has given me a wealth fascinating experiences.

In 1979 we planted about 10 hectares of cacao; the stuff chocolate is made from. The uphill side of the plantation was adjacent to a rainforest and the lowland side bordered pasture and rice fields. After six years, just when the plantation reached full production, the price of cacao plummeted to the point that cacao beans cost more to harvest than they were worth on the open market. We abandoned the cacao and soon the white-faced capuchin monkeys from the rainforest moved in and added cacao to their already hodgepodge diet. They don't eat the cacao seeds, rather they slurp the sweet-and-sour tasting syrup coating the seeds. That story is told in my first book, *Monkeys Are Made of Chocolate*.

Prior to the arrival of the monkeys, variegated squirrels were the only wild animals that ate our cacao. They didn't eat much, so we didn't worry about them. When the monkeys began frequenting the plantations on a regular basis, the squirrels disappeared. We could see them at the ecolodge and in an area that was planted to teak trees, but they were absent from the cacao plantations and forested areas.

Beginning in 1987 Hacienda Barú started offering ecological hikes with me as the only guide at first. Later, as demand grew, we hired more guides. None of us were biologists, but I had lived near the rainforest for 20 years, and all the others had practically grown up in the jungle. We all studied a lot so that we would be able to answer visitors' questions.

One of the guides learned that capuchin monkeys eat baby white-nosed coatis. They don't mess with the adults, but raid their nests in the tops of rainforest trees and eat the tiny helpless newborns. Squirrel nests are similar to those of coatis, and baby squirrels are even more helpless than baby coatis, so we assumed that they ate baby squirrels too. We later learned that the monkeys would also chase the squirrels and knock them to the ground, but that unless the monkeys were incredibly lucky it was pretty hard to kill and eat an adult variegated squirrel whose nutcracking incisors are sharp. All the squirrels moved out of the cacao and lived only in places where monkeys were absent and life was easier. We later learned that capuchin monkeys eat iguanas or just about anything else they can catch and kill.

Hacienda Barú Lodge is bordered by cacao on one side and secondary forest on two. These are all places where monkeys are frequently seen. On the fourth side is a narrow access road to Barú Beach. Across that road is a former rice field that, in the mid-1980s was slated for development at some time in the indefinite future. We divided it into 24 lots with teak trees on half of each lot and mango and citrus trees on the other half. We figured that the lots could be marketed in 20 to 30 years, once there was decent access and infrastructure into the area. They would be attractive to prospective buyers because the teak could be harvested for construction of a home and there would be producing fruit trees near the home. Monkeys weren't interested in the area because it was on the other side of the beach road, which bordered pasture and teak trees.

Things don't always work out as planned. Today, 35 years later, none of the lots have been sold and no houses have been built, 90 percent of the teak has been cut for lumber to build Hacienda Barú Lodge, and a variety of native trees have been planted where the teak used to be. The mango trees are big and beautiful and produce loads of mangos. In 2018, we strung a thick rope bridge over the road to the beach so the monkeys could access a couple of mango trees, and on occasion I saw them come to the ground and run across a 20-meter gap to get to a third mango tree. They became very fond of mangos.

All this time the native trees planted within the former teak plantation were getting bigger and bigger. One day in 2021 during mango season I went on a late afternoon hike to get some exercise and see if I could find a couple of mangos. But I found something even more exciting. A troop of about a dozen monkeys were playing around in a couple of the native trees which were mixed in with the few teak trees that hadn't yet been cut. For me, seeing the monkeys in that location for the first time was a special treat.

Three days later I went on another late afternoon mango-picking hike and decided to go to a tree I had seen recently replete with nearly ripe fruit. It was a good walk to get there, and I could walk home along the edge of the old teak plantation and hopefully see some monkeys. I was in for a big surprise. The mango tree was full of monkeys who were picking mangos like crazy. They took one or two bites out of each one and then threw it on the ground.

I stood there watching for a minute before walking under the

tree to try and salvage a mango or two. The monkeys got furious, screaming and throwing mangos at me. They were so intent on driving me away from their tree that they didn't bother to bite into the mangos before throwing them, and I was able to grab a couple of whole, unbitten ones and escape the bombardment with nothing worse than a bruise on my left arm.

I went home grinning from ear to ear with a beautiful, ripe mango in each hand. When I showed Diane and told her about the bombardment, she laughed and said, "There you go. Another book, *Monkeys Are Made of Mangos.*"

2. A Ball of Fluff Called Equinox

A Rescue? A Kidnapping?

Equinox is a Latin word meaning "equal night." Twice each year, when the sun shines directly over the equator, day and night are the same length all over the globe. The vernal equinox, which marks the beginning of spring in the northern hemisphere, falls around March 21; the autumnal equinox, marking the beginning of fall, is around September 22.

The spring equinox in the year 2009 fell on March 20, and that was the day a little ball of fluff came walking into my office at 6:30 in the evening just as I was thinking about closing up and going home. It was obviously a very young owl, nearly ready to fledge, but not yet able to fly. We had heard an owl calling near the office on quite a few occasions, and I surmised that this youngster had fallen out of its nest, and that its chances of getting back were almost nil.

Normally I believe it is best to let Mother Nature deal with her own creations. Her ways may sometimes seem harsh to us, but if a baby is separated from its mother and destined to die, nature has a good reason. I don't feel we should interfere in this process. But this cute little ball of fluff was too much for me. I couldn't bring myself to shoo it out of the office and into the cruel world. Instead, I called Diane, who totally disagrees with my philosophy about survival of the fittest, and asked if she would be interested in trying to rescue a helpless baby owl. "Do you have to ask?" she replied. "Bring him over."

By the time I arrived home Diane had already called an ornithologist friend and had figured out what to feed the fledgling, mostly small chunks of meat. Owls are 100 percent carnivorous and eat only animal protein. Though we had no idea what sex the owl was, we somehow assumed that it was a "he" and began calling him "Equinox." Later that was shortened to Iggy.

With help from Charlie, the ornithologist, we determined that Iggy was a screech owl of which there are three possibilities, vermiculated screech owl, Pacific screech owl, and tropical screech owl. All are small, less than 25 centimeters tall, and all are found in this region. We wouldn't know for sure until Iggy got his adult plumage.

Charlie explained that chunks of meat were not enough. Young owls eat mostly insects that their parents bring to them, and which they later learn to hunt for themselves. He said that wing and body parts had plenty of nutrients that weren't available in raw meat, and that these are essential for the young owl's digestion and development. We began capturing crickets and grasshoppers which we held out to Iggy until he grabbed them with his claw and stuffed into his mouth. Within a couple of weeks we were turning them loose in his cage, and he hunted and killed them by himself. Our overall objective was to teach Iggy to survive in nature and liberate him into the wild.

Charlie said Iggy needed mice in his diet too, with hair, skin, bones and all. He said that the parent birds would sever chunks of mice with their beaks and feed them to the young. Diane became a regular customer at the only pet shop in San Isidro that sold white mice. Before long she had a couple of cages full of mice and was raising her own. Charlie said that the most humane way of sacrificing the mice was to freeze them to death, so we put several at a time in a plastic jar and stuck it in the freezer. Later we cut them into slices, which I called mouse chops, and which soon became Iggy's favorite food. Within a month Iggy was ready to start killing his own mice. At first, he was pretty clumsy and the mice suffered the consequences of his inexperience, but after a few tries he learned to dispatch them with a quick bite to back of the neck. By this time, he had all of his flight feathers and easily maneuvered around his large cage.

During all this time we were getting to know Iggy and he was getting to know us. Owls have so many facial expressions and mannerisms that we tend to imagine that we know what emotion they are experiencing...happiness, sadness, fear, indifference, curiosity, disapproval, or anger. In reality, we probably don't have any idea what is going on in the owl's head. One thing was clear: he liked Diane. He loved to sit on her shoulder and preen her hair, as if it was a coat of feathers.

As he grew older his repertoire of sounds increased. As the name "screech owl" implies, they can emit a pretty loud screech. The call is, however, in no way unpleasant, unless, of course, it comes from nearby at 2:00 a.m. when you are sound asleep. At first his voice reminded me of an adolescent boy, but within a month he was calling just fine. In addition to the screech, Iggy made a number of other calls including a twittering sound that seemed to mean he was happy.

Diane eventually agreed that it was time to begin the liberation process. We started by taking Iggy outside and setting him down on a tree limb. He looked around a lot, swiveling his head more than 180 degrees, but didn't attempt to fly away. Diane stayed nearby, and he always jumped back on her shoulder when he'd had enough of the great outdoors.

After a couple of weeks of this Diane took him over to the butterfly garden, a large tent-like area covered with netting. He was plenty old enough to fly, but hadn't had much practice. His cage was a good size, but too small to permit extensive flying. The butterfly garden, at 16-by-16 meters, was a perfect place to practice. There were lots of posts, plants, and guy wires that had to be avoided, making the learning experience even more realistic. At first Diane only left him there during the day and took him home at night. Later he stayed there day and night. She mounted a box up high where he could take shelter when it rained. About every third day she would take him a live mouse and put it in a large open topped enclosure on the ground. The mouse could run around but couldn't get out of the enclosure. This hunting practice was crucial if he were ever to live in the wild and fend for himself.

A group from the Birding Club of Costa Rica and their ornithologist guide saw him in the butterfly garden and positively identified him as a tropical screech owl (*Megascops choliba*), a species which is found from Brazil to Costa Rica and is fairly common in our region.

The next step was to give Iggy total freedom of movement. Iggy was free to fly anywhere he liked, but there was a place in our garden where he knew that Diane placed a few chunks of meat at the same time every afternoon. She moved the mouse enclosure over near the same place and gave him a mouse now and then until the mice were all gone. As time passed Iggy came by less and less for free handouts, and eventually gave it up altogether.

At some point Iggy decided that he would sleep in our bedroom during the day. Diane placed a cardboard box up high in a corner where he would have a little protection. It was during this period that I learned how truly amazing owls are. Iggy could be sitting in the box with his eyes closed, and a cat would quietly tip-toe into the bedroom on the ceramic floor without making a sound. In an instant the owl was on full alert. It only took a couple of dive bombing attacks on the cats to teach them to stay out of the bedroom. Iggy always left about dark. Sometimes he would come into the house to hunt insects and sometimes he stayed out all night. He usually came flying into his box around 4:30 a.m., occasionally announcing his arrival with a loud screech.

When people try to raise wild animals, the animals never become domesticated; they only lose their fear of humans. As they mature, are better able to fend for themselves, and lose their dependence on humans, instincts take over and the wildness dominates their behavior. This happened with Iggy. There were days he didn't come to the bedroom to roost, and when he did, he started showing aggressive behavior toward humans. After several dive bombing attacks on Ana, the girl who helps Diane around the house, and one on Diane, we took the box down. Iggy got the idea right away and found a place outside where he could sleep during the day. He often came into the house at night to hunt insects, occasionally screeching in the wee hours of the morning, but all of the aggression was gone. Perhaps, to his way of thinking, the attacks were a defense of his roosting box.

The biologist suggested that we build a nesting box and gave Diane the design. The box was 35 by 50 centimeters and had a hole the size of a baseball in one wall, just large enough for an owl to slip through. We mounted it out on the porch. By this time it was early in the year 2010, and it would take a year before Iggy showed any interest in the box.

During that year we heard him call almost every night, sometimes from out in the garden and at others from our front porch, where the nesting box was located. Eventually we got the idea that there were two different owls calling, but there was still no sign of interest in the box. On many mornings there was a telltale white spot or two on the living room floor attesting Iggy's nocturnal visits. Once in a while he would come inside to hunt in the evening while we were reading or watching TV, but there was no physical contact. Iggy was definitely wild, albeit with an affinity for our house, both inside and out. Around Christmas of 2010 was the first time we saw two owls together. Later we saw signs that they were investigating the nesting box.

About five o'clock on the evening of March 18, 2011, I walked into the storeroom at the office and came face to face with another ball of fluff, a dead ringer for Iggy when he first appeared, and practically in the same place. This one was on a chair. Vanessa, the receptionist, spotted a second ball of fluff on one of the shelves, and the following day a third one appeared. The day after that the mother owl started appearing with the fledglings in the late afternoon.

It was then that everything became clear. We hadn't really rescued Iggy at all on the equinox of 2009, rather we had kidnapped him from his mother, and taken him on a strange adventure that few owls will ever experience. The mother owl had made her nest in a secluded cubbyhole in the ceiling of the office storeroom. When the chicks started to fledge was when we noticed them on the floor, the furniture, and the shelves of the store room. One evening we saw all three of them sitting in a tree outside the office with their mother just at dusk. That was the last time we ever saw them.

By this time, back over at the house we were seeing lots of activity on our front porch around the nesting box. With tropical screech owls, the male chooses the nesting site and claims it as part of his territory. The female decides which male she will mate with, in part, on the basis of the desirability of his nesting site. I like to think that his sexiness has something to do with it as well. Anyway, here is where things start losing clarity, because now we aren't sure if Iggy is a male who convinced a female to nest in his box, or if Iggy is a female who sweet-talked a male into accepting her choice for a nesting site. What is clear is that two owls, one of whom was most certainly Iggy, nested in the box.

On Sunday, July 3, 2011, I got up as usual at 5:00 a.m., opened the doors to the porch, and came face to face with a ball of fluff standing on an old couch we have out there. It was more mature than Iggy when he/she first appeared in the office. This ball of fluff had enough feathers that it could fly up into the rafters where it was joined by an adult owl, who may or may not have been Iggy. Every day they moved to a different place, but always on the porch.

When Diane walked out onto the porch, the adult twittered at her like Iggy always did. At night the adult owl went out, hunted and took insects back to the baby, who I call junior. On the sixth night they both disappeared. I imagine that the adult is carrying out the education of junior. We were interested to see if the adult owl would teach junior to hunt in the house in addition to hunting in the wild. It would appear that by kidnapping Iggy back in 2009 we had created a generation of owls with an affinity for human homes. This custom, in part, was begun by Iggy's father, who charmed his mother and convinced her to nest in the office storeroom. Had she nested out in nature where most owls do, this story would never have been told.

Iggy on Diane's shoulder.

Epilogue: The year is now 2023, and tropical screech owls still come into the house to hunt insects, mainly crickets. In one or two nights of hunting they will kill all the crickets and won't be seen for three or four months. By that time, we will be seeing crickets again and the owls will return to hunt them.

3. Unlikely Partners

Quit Chopping the Weeds and the Jungle Returns with a Vengeance

Environmentalists and people in the real estate business don't usually like each other very much. The former tend to blame the latter for environmental destruction while the latter accuse the former of obstructing development and progress. In spite of this, on a strip of coastal land in southern Costa Rica 80 kilometers long and 15 kilometers wide, nothing short of an environmental miracle has been accomplished by these two actors working separately and each doing their own thing. The area is called the Path of the Tapir Biological Corridor (PTBC) and its landscape has been transformed from one of pastures and farmland to lush tropical forest replete with wildlife. It is one of the few places on our planet where biodiversity is increasing and has been for the last 35 years.

It all began in 1987 when the Association of Friends of Nature (ASANA) was founded by a group of local people who were alarmed at the rapid increase in illegal hunting in the area around

Dominical. With the paving of what was formerly a jeep trail, many hunting enthusiasts from other places flocked into the area and proceeded to kill the diminishing wildlife. ASANA was successful in combating the illegal hunting, and the organization grew and expanded its area of influence.

In 1990 the Path of the Tapir Biological Corridor was initiated by ASANA. At that time about 80 percent of the area consisted of farms and ranches, and a mere 20 percent of the original rainforest that had once covered the entire area remained intact. Small parcels of forest here and there and several large, forested areas, such as those on Hacienda Barú, Rafiki, and Rancho La Merced, prevailed mainly on the steepest parts. The goal of the PTBC was to connect these forest fragments by rewilding (restoring natural vegetation) the land between them, thus creating a green corridor. It was named the Path of the Tapir because it was a place where tapirs had once

Tapirs captured on trail cameras.

been abundant, but were finally extinguished by hunters in the mid-1950s. Nevertheless, these large mammals were still found in the Los Santos Forest Reserve to the north and the Osa Peninsula to the south.

ASANA accomplished its objectives through environmental education, working with government programs which provide incentives for forest protection and regeneration, denouncing illegal hunting and destruction of forest, and working with those who promote ecological tourism. Tapirs haven't yet returned to begin reproducing in the corridor, but there have been numerous sightings, and most naturalists believe that within the next decade we will see their definite return.

Nevertheless the project has been enormously successful. When I first visited Hacienda Barú in 1972 only the white-faced capuchin monkeys were found here. Today we have all four Costa Rican species of monkeys. Sightings of large predators, such as pumas and ocelots, are not unusual. Numerous new species of birds including the scarlet macaws, crested guans, oropendolas, green ibis and others now frequent the area. Ecological tourism has become an important business in the PTBC and is attracting thousands of tourists every year and generating employment for hundreds of local people.

As much as I wish I could say that the phenomenal success of the PTBC was all the doing of ASANA, I would not be telling the truth. The late 1980s brought the initial stages of a real estate boom. People started arriving in the area looking for land. Most potential buyers were foreigners looking for a beautiful place to build a vacation home, or a bed-and-breakfast place. They definitely weren't interested in raising cattle or farming rice.

As land values appreciated sellers began noticing that the properties with the most natural vegetation sold best. The buyers loved seeing monkeys and toucans and other wildlife. They also loved those hilltops with the fantastic ocean views. It had never occurred to the original owners that anyone would value a steep, hard-to-access location just because you could see the ocean. Property owners stopped chopping the weeds in their pastures and allowed the secondary forest to regenerate. A professional forester once told me: "In Costa Rica, when you quit chopping the weeds, the jungle comes back with a vengeance." And that's what it did.

At the same time bulldozers started cutting roads up hillsides to access those ocean-view sites and the environmentalists panicked. But the returning jungle soon overshadowed the red dirt exposed by the bulldozers. The panic gradually diminished and eventually subsided. The new owners loved the wild animals and birds, and many joined the environmental movement. Many others didn't join an organization but contributed to the greening of the PTBC by allowing natural vegetation to return and by protecting their properties from hunters.

Hacienda Barú and Barú River, 1972

Open areas between forest parcels disappeared and the jungle merged into one big swath of tropical vegetation that permitted free movement of wildlife from one place to another along the width and breadth of the corridor. It was wonderful watching it happen, but few really understood the part played by the real estate boom, not even those directly involved in the process.

If anyone in the environmental movement understood the part played by the mass change of ownership in the area, they certainly didn't publicize the fact. In all fairness, the real estate boom didn't directly contribute to the greening, rather it facilitated a change of land stewards, from those who cut down the forest and planted pasture and crops to those who quit chopping their weeds and allowed the jungle to return—with a vengeance.

To my knowledge no measurements exist of the natural cover in 1985, the peak of the deforestation era and today in 2023. Looking at old and new aerial photos I estimate that around 80 percent of the land area had been deforested by 1985 and was being used for livestock or farming. Today the ratio of secondary and primary forest to open areas has reversed. About 80 percent of the land is covered with natural habitat. Thanks to these strange partners, who didn't even know they were partners, the PTBC is very close to accomplishing its goal.

Imagine what would have happened had they consciously worked together from the beginning.

First photo: Hacienda Barú and Barú River, 2015

4. Battling Africanized Bees

My Struggle with the "Most Successful Biologically Invasive Species Ever to Plague the Western Hemisphere"

Hacienda Barú Lodge has been in operation since 1990, and I thought I had experienced all of the problems that could possibly confront a hotel manager. That was until one warm March evening in 2012 when a swarm of Africanized bees invaded one of our guest rooms. The guests had returned to their room about 6:30 p.m. and found it full of bees. The first hotel employee they encountered was the guard who they told about the problem. He went to see for himself and called the office on his radio.

"There are a lot of bees in #24" he told me.

"How many is a lot?" I asked, "10, 20, 50?"

"Oh no, lots more. There are thousands of bees in #24."

"I'll be right there," I said, and headed out the door.

When I walked into the room, I couldn't believe my eyes. Hanging from a wall lamp directly above the bed was an elongated mass of bees shaped kind of like an American football, but twice as big. The guard hadn't exaggerated. There were literally thousands of bees clinging together in one bunch and others flying around the room. I knew enough about their life cycle to know that the new queen was at the center of that mass which is, more correctly, called a swarm. Their intention was to create a hive in room #24. I, of course, was determined not to let that happen.

Fortunately, the swarm was calm, probably because of the late hour. I knew that the fire department was usually available to deal with problems with bees, but they would have to come all the way from Quepos. I also knew that they spray soapy water on the hive or swarm. Apparently, the soap sticks to the bee's wings and immediately incapacitates them. I found a sprayer with a two-gallon tank, dumped in a double handful of laundry detergent, filled it with water, and shook it up. Back at the room I covered the bed with plastic and began spraying the soapy water in the air and directly on the swarm. Flying bees fell to the floor where they fluttered around and eventually died. Bees peeled off the swarm and fell to the floor.

I had to refill the sprayer once, but finally the oblong mass was gone and thousands of dead and dying bees were all over the floor. After closing the office at seven o'clock that evening, my assistant Shirley came to help. We swept up a small waste basket full of bees and swept all the water out the door. Had we had another room available I would have relocated the guests, but we were fully booked. They were very understanding and said they didn't mind sleeping in the same room. All the bees were gone. Cleaning up wasn't completely done, but we cleaned the floors and changed the bedding. I gave a big sigh of relief. Problem solved.

Or so I thought. The next morning there were bees in rooms 22, 23, 24, 25 and 26 but less than a dozen in each room. They weren't aggressive and seemed disoriented, but there were enough of them to upset the guests. We continued spraying the soapy water inside and outside of the rooms with the same results, the bees died, and later in the day or the following day the rooms again had a few bees inside, and others hanging around outside. We couldn't understand where they were getting into the rooms which had always appeared to be mosquito proof. We tried sealing every tiny little crack or opening where a bee might squeeze through. But they kept returning on a daily basis. Eventually, as the rainy season settled in, the bees quit coming. I was elated, but cautious. I doubted that this was the end of the story.

As it turned out it was the end of the story for that year. The following dry season, a year later, the bees were back. I consulted with the fire department in Quepos and learned that there was probably a large hive in the area, which the fireman referred to as a "mother hive." In February or March every year the larger hives produce a new queen which leaves the hive with a large group of worker bees. These are referred to as a swarm, and their purpose is to start a new hive. That's what happened in room #24 a year earlier. After we killed the queen that first night, the remaining workers stayed in the area, but having no hive or queen to defend, their aggressiveness was gone. The problem was that the first night, when the bees swarmed in room #24, they left an odor in the room that is undetectable to humans but will continue to attract bees for years.

Knowing all of that was very enlightening but didn't help us solve the problem at hand. I told Olman, our maintenance man, to check with the house cleaning crew every day at mid-morning and, if there were any bees, to deal with them. After a couple of weeks, I realized that we hadn't received any guest complaints about bees for four or five days. I checked with house cleaning who informed me that they hadn't seen any. I was puzzled by the situation. The year before the problem had persisted into June, and we were only in February.

One day I made a comment to Olman, about how odd it was that the bees had left so suddenly. "Nothing odd about it," exclaimed Olman, "you told me to deal with the problem, so I did. I know you don't like pesticides, and neither do I, but this situation was serious. Soapy water wasn't doing the trick. One day when I was in the farm supply store, I asked the clerk if they had a pesticide that would kill bees and explained our problem to him. He sold me a product and told me to spray it on the outside of the rooms. 'It will get rid of your bees,' he told me. So that's what I have been doing. Once a week I spray the outside walls of the rooms, and the bees don't come around anymore."

During my many years in Costa Rica, I've come to admire the intelligence and problem-solving ability of campesinos, and this incident served to strengthen that respect. It was a great example of the simplest solution being the best; if what you are doing doesn't work, do something else.

Our experience with what biologists call Africanized bees began years earlier. During the 1950s some scientists in Brazil imported about 30 queen bees from an extremely aggressive African species in the hopes of increasing honey production in Brazil. The problem began when 26 swarms escaped into the wild in 1957. They crossed with local bees thus producing hybrids that were more aggressive than the local bees and less aggressive than the African bees.

By the time they reached Costa Rica in the mid-1980s, they were considerably less aggressive than the original African species, but considerably more aggressive than our local honeybees. Africanized bees have a larger defensive area around the hive, react more quickly to any perceived threat, react in larger numbers, and chase the victim farther. Over 1,000 humans and many farm animals in the Americas have been killed by them.

The US Department of Agriculture was concerned that the Africanized bees would migrate as far as the United States. They decided to try and stop them in Costa Rica, the narrowest point in the Central American isthmus. They trained a group Costa Rican Ministry of Agriculture (MAG) employees and all the fire departments to combat them, providing the necessary resources for salaries, vehicles, and equipment. For a couple of months in 1984 MAG employees and USDA personnel were a common sight. They dealt with several hives on Hacienda Barú. They used diesel rather than soapy water. It worked well but quickly destroyed the rubber seals in our sprayers.

One day the USDA received word the bees had been reported in El Salvador. In 1985 they appeared in the San Joaquin Valley in southern California. In Costa Rica, the Africanized bees have taken over completely. There are no more of the original local honeybees left here. (An interesting side note: most of the local honeybees had originally been of an Italian species which had been selectively bred for honey production and their gentle disposition.)

Beginning in January of 2014, scarlet macaws began visiting Hacienda Barú and the surrounding area on a regular basis. In mid-July of that year the macaws' behavior suggested they might be scouting out nesting sites. We decided to put up nesting boxes in four trees that the big red birds frequented and see if we could help them out with their endeavor. Many types of boxes are described on Internet, but we chose one made of wood and measuring 35 x 35 x 70 cm. All were mounted about 25 meters above the ground. We waited for the macaws to move in, but that never happened. Not only did they not nest in the boxes, but they quit visiting the trees. We thought that maybe parrots or parakeets would use them, but no such luck. The boxes just hung there, empty, we thought.

In February 2015 one of our guides focused his telescope on the entrance hole to a nesting box to see if anything at all was inside. What he saw astounded everyone. The box was full of bees. No wonder the macaws weren't interested in them and even stopped visiting the trees where they hung. A quick check confirmed that all four boxes contained active hives of bees. With that many hives in such a small area they would all be competing for locations suitable for new hives and could be expected to invade any natural cavity including houses and hotel rooms.

Again, we were faced with a bee problem, a much bigger one complicated by the fact that the hives were located 25 meters above the ground. I drove to Quepos to talk to the firemen. I learned that they had no means of dealing with a hive that far off the ground. I explained that we had three guides who frequently guide people into the treetops on a tour we call Tree Climbing. Any of them could easily climb up to the hives. We discussed several different plans, and consulted with the guides who would have to carry them out.

In the end the fire department lent us a protective suit which one of the guides could wear while climbing up to the box. The climber could carry a two-gallon sprayer with a strong pesticide that would kill the bees instantly. Once the bees were all dead, the box could be dropped to the ground. The fire department preferred that we use soapy water, but I decided on the faster acting pesticide out of concern for the climber's safety.

It went like clockwork, Carlos and Pedro first put a rope over a branch near the hive closest to the lodge. Carlos then put on the suit, hung the sprayer over his shoulder and a small wrench from his belt, and began the long climb. The suit was hot even though it was 6:30 a.m. There were a few bees flying around the outside of the hive which attacked him a couple of meters before he reached the box. He sprayed them and quickly closed the distance. Once close to the box he stuck the nozzle through the four-inch hole and sprayed the inside. All the activity around the hive quickly diminished and then ceased. Every single bee, including the queen, was dead. Finally, he removed the cable clamps and let the box drop to the ground, later to be burned.

We had to limit our activity to one hive per day, early in the morning. As the sun climbed higher in the sky and began to warm things up, the bees became much more aggressive and attacked the men on the ground while they were trying to get the rope in the tree. Carlos did the first two trees, Ronald did the third, and Carlos did the last one.

We are very grateful to the Quepos Fire Department for the help they have given us. They are true professionals and very willing to help the public. Thanks, guys.

As I write these words Carlos has just finished with the last box. We have won another battle against the Africanized bees, but somehow, I don't think the war is over. We will just have to wait and see.

You may have heard that bees have been declared the most important living creatures on the planet. I've seen this news item at least five times on Facebook. I agree that bees are very important pollinators, but it would be difficult to declare any one class of living things as the most important. There were several different postings with differing headlines like: "Royal Geographic Society of London Declares That Bees Are the Most Important Living Being on the Planet," "World Watch Declares at Royal Geographic Society.... etc.," and "Scientists Declare That Bees.... etc." All the articles were vague about who actually made this declaration. All insinuated that it had been recent. They mainly focused on how important bees are and the horrible consequences to humanity if they were to become extinct. If bees are really the most important living beings on Earth, maybe we shouldn't kill them even if they are interfering with the macaws nesting.

But something just didn't sound right. My curiosity and skeptical nature got the best of me, and I started digging. Finally, I got right to the nitty gritty, a 2008 article in a British newspaper, *The Guardian*. Apparently, every year the Worldwatch Institute used to hold a debate at a meeting of the Royal Geographic Society. At the 2008 event five scientists debated over which was the most important living being on Earth: bats, bees, fungi, plankton, or primates. After the debate the audience voted, and bees came out the winner. What that really means is that the scientist who presented the case for bees was the best debater.

I agree that bees are very important to the well-being of the Planet. Nevertheless, there are around 25,000 species of bees in the world. The one that is plaguing the scarlet macaws is a rogue hybrid that is a danger to animals and humans. In my opinion we are just beginning to learn how damaging they have become to our tropical rainforest ecosystem. I feel certain that more information will come to light in the future, and that it won't be good.

For me there is no dilemma. Kill the Africanized bees wherever you find them.

5. Cool Cats

Sighting Felines in the Wild

Experiencing the sight of a wild feline is about the most exciting thing that can happen to a nature enthusiast in Costa Rica. Most humans share a fascination with cats, whether it be the tiny oncilla *(Leopardus tigrinus)*, the smallest cat in Costa Rica at two kilos, or the enormous Bengal tiger *(Panthera tigris)*, at 240 kg, the largest cat in the world. As they are naturally secretive and wary of humans, seeing a wild feline is not an everyday occurrence.

Though we have five feline species at Hacienda Barú. I have only seen one, the Jaguarundi *(Herpailurus yagouaroundi)*, which I have spotted on multiple occasions. One of those sightings was particularly memorable.

A neighbor who lived close to the lodge bought some young chicks to fatten for meat for the dinner table. They grew quickly and soon the roosters among them began to crow. I know of no sound more irritating than the crowing of an adolescent rooster. It seemed like there was a competition among them, and each day they started a little earlier and crowed a little louder. Soon the racket began shortly after midnight. Of course, our guests complained about it, but the neighbor, Ricardo, was uncooperative.

One day I saw two jaguarundis cross the driveway heading toward his chicken coop. Word soon got out that they had killed a couple of chickens. Knowing that Ricardo would try to kill the cats, I called the wildlife department and asked if a game warden could have a talk with him. One happened to be in the area and stopped by that same afternoon. He told Ricardo that if anything happened to the pair of beautiful black cats, they would know it was him, and that the fine would be enormous. Ricardo killed all the young roosters that same day and put them in the freezer.

I don't think that these beautiful black cats are seen more frequently than other felines because they are more numerous, rather it's because they are diurnal and therefore see more humans. As they have never been harmed by a human, they are less wary of them. Also jaguarundis spend most of their time on the ground. At 5 kg, the jaguarundi is the third largest of the five Hacienda Barú species.

Although I have never seen one, the ocelot *(Leopardus pardalis)* holds a special place in my heart. It is responsible for me becoming an environmentalist. In 1974, when Hacienda Barú was a cattle ranch, the cowboy killed an ocelot. It came around his wife´s chicken coop, his dogs chased it up a tree, and he shot it out of the tree. I had hunted deer and ducks in Colorado where I grew up and didn't have any problem with hunting, but the sight of that beautiful spotted cat's dead body laid out on the porch of the cowboy's house affected me deeply, to the point that I gave up hunting and began a long arduous struggle to stop illegal hunting on Hacienda Barú.

After many years and many helping hands, the result is the Hacienda Barú National Wildlife Refuge, where all the wildlife now feels much more secure than before. My wife Diane has seen two ocelots, all our guides and park rangers have seen at least one, and our guests sight them a couple of times each month on average. Both adults and kittens appear on our trail cameras at multiple locations.

I feel that the population is healthy and stable. At 12 kg the ocelot is the second largest feline on the hacienda.

According to *The Mammals of Costa Rica* by Mark Wainwright, the puma *(Puma concolor)* weighs around 50 kg, making it by far the largest feline at Hacienda Barú. Though the Costa Rican puma is slightly smaller than the North American cougar or mountain lion, they are all the same species. The two populations are pretty much isolated from one another; the northern variety feeds mostly on large game like deer and elk, while our pumas never kill anything larger than a deer and most of their diet consists of peccary, opossums, raccoons, coatis, and even smaller animals. For that reason, natural selection has favored larger cats in the north.

Pumas were abundant for the first part of the last century. They were all killed off in the 1940s and '50s by farmers and ranchers and have only recently returned. The first sighting on Hacienda Barú was in 2009 by a couple who were walking on one of our trails and came face-to-face with a pair of young pumas. Park rangers sometimes see them near the beach during turtle season. They don't bother the turtles, but hunt the raccoons, coatis and dogs that dig up the turtle eggs. The population has grown to the point that we capture two or three photos of pumas on the trail cameras each month.

A margay *(Leopardus wiedii)* has never appeared on one of our trail cameras, and there have only been two sightings on Hacienda Barú, both by our guides and the guests they were guiding. Most of the margay's time is spent in the treetops where they lead a solitary nocturnal life. One source says that they only come to the ground to defecate and move across a gap to another tree. At 3.5 kilos, this species is the second smallest on Hacienda Barú.

The oncilla *(Leopardus tigrinus)*, sometimes called the "little tiger cat," is even smaller than the margay, weighing a meager two kilos. None of our staff has ever seen one, and *The Mammals of Costa Rica* says that they aren't found in this part of the country. Two sightings, both by biologists at different locations and about a year apart, give us confidence that they do exist on the refuge. Along with the rest of the spotted wild felines, they were hunted heavily in the 1960s for the fur trade. It takes 24 oncilla pelts to make a fur coat. Fortunately, there is no longer a market for fur coats, and hunting pressure has diminished to almost zero.

Prior to 1940 jaguars *(Panthera onca)* were often seen at Hacienda Barú and throughout the area that is now known as the Path of the Tapir Biological Corridor. As their territory was invaded by humans and their habitat destroyed for farming and ranching, they turned to domestic animals for food. Pigs seemed to be their favorite, but calves, chickens and dogs were also on their menu. The last jaguar in this area was killed in the mid-1950s. Hopefully they will return across the biological corridor one day as more habitat is restored and people learn to appreciate our wild heritage.

I have never seen a jaguar in Costa Rica, but I was fortunate enough to visit the Pantanal National Park in Brazil with the express purpose of seeing one in the wild. My daughter Natalie sighted the first one a second or two before me. "Oh Daddy, it is huge!" she exclaimed. And they really are huge. At around 80 kg, they are considerably bigger than a puma and are undisputedly the largest feline in Costa Rica.

6. The Goldwalker

Walking the Jungles of the Osa for 26 Years

If you saw a small Irish man with a beard and bushy eyebrows who lives in the forest, loves the color green, and has a passion for gold, you might think he was a leprechaun. But if that forest happened to be on the Osa Peninsula in Costa Rica, more than likely it would be my good friend Patrick J. O'Connell, commonly known as "The Goldwalker." Unlike the leprechaun he is not the keeper of the pot of gold at the end of the rainbow, but he has lived the most extraordinary life of anyone I've ever known, walking from one end of the Osa to the other for 26 years buying gold.

In 1964 Patrick and three friends drove to Costa Rica in a 1958 Dodge panel truck. All were veterans of the Korean war and out for adventure. They came to Costa Rica for the hunting. For three years he stayed in the Talamanca mountain range in southern Costa Rica near a place called Potrero Grande. Then Patrick moved to the Osa Peninsula where the hunting was spectacular.

As it is today, the Osa of the 1960s was one of the most biodiverse areas on the planet, but that is where the similarity ends. At that time there was no national park, no tourism, no roads, and most of the residents were fugitives from the law. After the civil war of 1948, many of the soldiers from the losing side fled to the Osa to escape execution. Additionally, criminals of all sorts took advantage of the remoteness of the peninsula and went there to escape punishment. It was a place of exile. No criminal act was terrible enough to bring the police to the peninsula. If someone was killed in a barroom brawl, the others just dragged the body outside and the next morning someone would bury it. The authorities didn't care.

When I first visited the Osa in 1972 we flew into a short gravel air strip in Puerto Jiménez. The small town consisted of a couple of bars, a pulperia (general store) and even a gas station. All of the fuel came by barge in 55-gallon drums. There were a few cattle ranches along the coast near the town. After finishing our business there, we hired a boat to take us to Golfito. The captain told us that occasionally one of the exiles would delude himself into thinking that the police had forgotten about him and try to return to the mainland. What he didn't know was that a couple of policemen with a list and photos waited for all the boats, and always captured the unlucky fugitive.

Since pre-Columbian times, the Osa has been known as a rich source of gold. Most of the inhabitants at the time of Patrick's arrival made a meager living extracting the precious yellow metal from the rivers, while living off the land. At first all Patrick did was hunt in the game-rich forests of the peninsula and eventually got to know the area like the palm of his hand. He also befriended most of the miners, men with names like Chente, Mico Blanco, Zorro de Agua, Bomba, Chontales, Mico Negro, Carafila, Juan Bravo, Cleaver, Ampia brothers, and Omar. Almost all had at one time murdered someone and decided to live a life of exile rather than go to prison. Living in the jungles of Osa and extracting gold from the rivers was a hard life, but these men preferred it to living behind bars.

Chente, a gold miner in Costa Rica (1970)

In the late 1960s the options for selling gold were limited. The international price of gold was only $35 per Troy ounce, and the few buyers who did come around gave the miners a pittance for their hard-earned gold.

In the early 1970s the international gold markets got an enormous boost. First, the price of gold in the US was deregulated and then all sanctions on American citizens owning or trading gold were lifted; the precious metal became a commodity just like corn, cotton or coffee. From that time forward it was listed on the New York Commodities Exchange. The price of the Osa's only product began an unprecedented ascent reaching a value of $850 per Troy ounce by 1980. (Gold is measured in Troy ounces. One Troy ounce is 30.1035 grams, while the commonly used Avoirdupois ounce is only 28.3495 grams.)

Patrick began his life as a gold buyer about the same time the price began its historic rise, and everyone thought that he was the one who made it happen. They also thought that if anything happened to him, the price would return to its former low level. From that point onward his safety was their concern. He walked everywhere there were miners, paid a fair price for their gold, never cheated anyone, and taught them how to find out the international price of gold by listening to the daily market reports on a portable short-wave radio. Most of them became his friends.

Everyone knew that Patrick carried large sums of cash and gold in his backpack, and he knew that it was a big temptation for some of the miners. He lived by several simple rules designed to keep him alive. He never told anyone where he was going next, and if asked he lied. In the forest he constantly watched and listened and felt confident that he could evade or out walk anyone who tried to follow him. He knew all the trails well and the best places to hide or outsmart someone who might follow. On the beach, where there was no place to hide, he only walked at night and never used a flashlight.

Even though the miners were receiving much more money for their gold, most weren't living any better than before. The majority of their earnings went for booze. There were cantinas in Puerto Jimenez, Rio Oro, Carate and in several other locations around the Osa. If someone found a big nugget and sold it for a small fortune, the whole bundle was spent on a big party that lasted until all the money was gone. Then they would all go back to mining until the next big windfall.

One of the most famous cantinas, located in Carate, was owned by a very tough lady named Doña Berta, who also happened to be Patrick's mother-in-law. Everyone went there to drink, but seldom caused problems for fear of Doña Berta's wrath, and her shotgun. One notable exception was when Juan Bravo shot and killed Chino Ahoi while drinking in the cantina. There was bad blood between the two, but Chino figured Juan Bravo wouldn't cause any trouble at Doña Berta's. He was dead wrong.

Juan Bravo was rumored to have killed several other miners on the Osa, but nobody knew for sure. Patrick eventually built a pulpería (general store) and bar near the Rio Tigre, and his wife Roxana managed it. On several occasions Juan Bravo walked into the bar, headed straight for Patrick, pulled out his revolver, pointed it at Patrick, and said: "Here. Take care of this for me. I'm gonna go drink, and the devil gets in my head when I drink." Like the other miners Juan Bravo never talked about his past, but it was commonly believed that his father was very religious, and he had been raised in a strict religious household.

Patrick has enough stories to fill several books, but the one that seems to upset him in the telling is about a money lender named Don Transito who apparently had a heart attack while bathing in a stream.

Some miners found him there either unconscious or dead. Pat walked up shortly thereafter, and they had Don Transito laid out on a wooden platform with his hands tied together and his jaw tied shut. When asked what happened one of the men said they had found his body down by the stream and were getting ready to bury him.

Patrick looked at the body and noticed the fingers were twitching. "I don't think he's dead," said Patrick. "Look, he's still moving."

"Oh, he's dead alright," said the miner with a look that told Patrick that any further comment would be unwelcome. Patrick looked at the three miners, all with machetes at their sides, and all of whom owed money to Don Transito. He turned and walked away.

One of the most violent incidences to take place on the Osa was perpetrated by a miner named Cleaver in the mid-1980s. By that time there was a crude road to the Osa and an occasional visit from the police. Cleaver was from Guanacaste and rumor had it that he had killed his wife and fled to the Osa. One night he was sitting in a bar drinking when another miner walked over, picked up Cleaver's drink, drained it, and sat down. Cleaver stood up pulled out his machete, made a deep cut on one of the man's shoulders and then brought the big knife down hard on his thigh, cutting it all the way to the bone. Blood gushed out of the cut, and the man quickly keeled over and died. Cleaver turned to the bar and asked if anyone else wanted to try to steal his drink. Then he left and went into hiding.

The police did a routine investigation but didn't mount a search for Cleaver. The captain told Patrick that if he or any of the other miners saw Cleaver that they should kill him and throw his body in the river. Sometime later, three of the dead man's relatives pursued Cleaver and got him cornered near Patrick's store. The dead man's brothers were unarmed as was Cleaver who looked crazed and half-starved and was throwing rocks at his pursuers. Patrick walked out of the store and pointed a revolver at Cleaver with every intention of killing him. Cleaver started crying, ran up to Patrick, threw his arms around him, and cried "Patrick, my friend."

"I just couldn't kill him," reminisced Patrick. "I told him to get out of there." Cleaver ran back into the jungle. Rumor has it that several days later the dead man's brothers caught up to him deep in the forest and killed him. He was never seen again.

Unlike the leprechaun, Patrick isn't the keeper of the pot of gold at the end of the rainbow, but during his life on the peninsula he bought and sold an enormous amount of gold and has seen a lot more. The biggest nugget he has ever seen weighed four pounds. The miner offered it to him, but he didn't like it. There was a lot of quartz mixed with the gold and it had an ugly shape. On another occasion he was offered a very nice one-kilo nugget, but didn't buy it either. The miner was showing it around and talking too much, and Patrick figured it was too risky.

The largest nugget he ever bought weighed 10 Troy ounces. In fact, he bought two of them that big. The prettiest nugget he every bought was in 1979, weighed 79 grams and cost 79 colones per gram. Large nuggets brought a premium price and Patrick cultivated a number of special clients who loved to buy them. But the bulk of the gold he bought was fine gold, sometimes called gold dust, and this was sold to the bank at the market rate.

During his hunting days Patrick killed three pumas and one jaguar. "I didn't want to kill the jaguar, but it was killing a friend's cattle, and he begged me to hunt it down." They chased the big cat with dogs, got it treed, and Patrick shot it out of the tree. Once Patrick started buying gold he only hunted for food for the table.

During all of his years of hunting and gold buying on the Osa Peninsula, Patrick saw many large snakes but was never bitten. Then, when in his 60s and on a camping trip with his son, he was bitten on the toe by two recently hatched terciopelo vipers at the same time. By the time they got him to the hospital in Golfito, six hours later, he was in critical condition. They were able to save his toe and foot, but the bite never completely healed. As of this writing, he needs a cane to walk at age 78, but even so you just can't keep him down. Now Patrick dedicates his free time to kayaking with his dog Kayak standing at the bow.

Patrick married Doña Berta's daughter Roxanna. Together they had four children. He made sure that all four had US passports and were dual citizens. All went to school in Indiana while staying with his family. He is very proud of all the kids, but especially the youngest girl, Rocki. She is an amateur boxer and Ringside World Champion at 101 pounds, pin weight.

In recent years Patrick formed a group called Golfito Veterans (GOVET) which includes American Legion, Veterans of Foreign Wars, Active Military members, and Korean Vets. The group has helped over 200 students in 30 schools in the Sierpe River school district. They have also donated 45 wheelchairs, numerous walkers, canes, and other supplies to the needy in the Golfito area. Their money is raised from yard sales, Super Bowl parties, and all the profits from the book *Goldwalker* by Scott Anderson, a fascinating biography that covers many of the years that Patrick lived on the Osa Peninsula.

First Photo: Jack Ewing with Patrick O'Connell (2017)

7. Quest for the Silky Anteater

The Golden Tennis Ball

Back in the 1980s the economy in this part of Costa Rica was still driven by rice farming and cattle ranching. Tourism was something we all knew would come later, but nobody had a clear picture of how it would work. In 1986 the government paved the road between San Isidro and Dominical, and the bridge across the Barú River was completed. Work on the Coastal Highway south of Dominical was in progress and proceeding slowly. A few tourists started filtering into the area, and cabins to accommodate them started appearing here and there. It was the very beginning of a new way of life for the rural community. In a few years it would come to replace farming and ranching as the most important economic activity in the area.

In 1987 we started doing two tours at Hacienda Barú. One was called the Rainforest Experience, and the other the Mangrove Walk. At that time Hacienda Barú had not yet been declared a National Wildlife Refuge but had a reputation as a private nature reserve. That first year we sold about $300 worth of tours. It wasn't much, but it was a start, and I really enjoyed guiding them. People liked the hikes, told others, and business increased. The second year we sold $1500 worth of tours, and the following year $6000. We were getting so busy that I couldn't guide all the tours, and we sometimes had to turn people away. The time had come to look for more guides.

We had a man we called the "forest guard" whose job it was to walk around the rainforest and stop hunters from killing the wildlife. His name was Juan Ramón, and he knew the trails, loved nature, and was well acquainted with the flora and fauna. His biggest drawback was that he didn't speak English, the language spoken by most of the clients. To make up for his lack of language skills I gave him a book called *Mammals of the Neotropical Rainforest*. When he spotted an animal Juan could show the visitor a picture of it in the book, along with the name and basic information about it. In those days most of the tourists that made it to out-of-the-way places like Dominical were pretty adventurous, and struggling with the language was just part of the adventure.

My biggest challenge was to convince Juan that guiding people through the rainforest was a worthwhile activity. He couldn't understand why anyone would pay to do something so ordinary as hike through a jungle. I tried to explain that the experience was something special for people from other countries. "If you won the lottery and took a vacation to New York," I told him, "you would happily pay to take a guided tour of the city. But someone who lived there would find it very ordinary." He kind of got the idea, but not really. It was going to take some more convincing.

That convincing came one day when Juan was guiding a group of four nature lovers on the Rainforest Experience. They came upon a three-toed sloth very near the ground, clearly visible and moving. I found out later that prior to that time Juan hadn't bothered to point out sloths to people. He couldn't imagine what anybody would find interesting about one of these plain-colored, lazy mammals. However, he made an exception that fateful day because the sloth was so obvious and it was moving.

To his surprise the people loved it. They got very excited, took lots of pictures, and one of the visitors gave Juan a $10 tip right on the spot. That was all it took. From that moment forward Juan never walked past any bird, mammal, reptile, or amphibian without pointing it out to the visitors. I believe that it wasn't so much the money as it was the idea that the rainforest he had thought to be so ordinary had real value for people from other countries. It was an eye opener for Juan, and his future guiding was filled with enthusiasm and a desire to show his guests every aspect of the jungle. That enthusiasm was contagious, and other workers who later learned to guide just seemed to absorb it.

Soon after we began taking people on hikes, we started putting together a species list. To begin, we put every bird and animal that any of the guides had ever seen on the list. The list grew rapidly at first but then slowed to a few birds and animals each year. Our first bilingual guide was an ornithologist named Jim Zook. His presence on the hacienda gave a really big boost to our bird list. Today, 30 years later, the list contains 369 species of birds, 85 mammals, 33 amphibians, and 63 species of reptiles.

In the early 1990s I became interested in a small rainforest mammal called the silky anteater (*Cyclopes didactylus*). According to all of the texts that I could find, it should exist in this region, but not only had none of our employees ever seen one, none of them even knew what it was. Neither did any of the neighbors. A biologist friend assured me that we most certainly had them on Hacienda Barú, but said the illusive little creatures spent most of their time in the tree tops and were rarely seen. Our best bet, he said, was to look closely at tangled loops of lianas and vines. I decided to make a competition amongst the three guides, Juan, Pedro, and Jim. But I needed a very special prize.

Before 1989 the only field guides available were *Birds of Panama* and *Birds of Mexico*; good books but incomplete for Costa Rica. The entire first printing of *A Guide to the Birds of Costa Rica* was purchased by the Audubon Society and sold only to their members. I doubt if a single copy made it to Costa Rica. With the third printing in 1992 a few copies of the book began to appear in the hands of Costa Rican guides and bird watchers, but the book was still a rare treasure and much coveted by bird lovers and people who worked in tourism all over the country. A friend of mine from the United States, who was a member of the Audubon Society, managed to acquire two copies and brought them to me on his next visit. I kept one, which I still have to this day, and offered the other as the prize in the quest for the silky anteater. The first guide to spot one would become the proud owner of this treasured book. It was the ideal incentive.

The silky anteater is the most charismatic animal I know. Its golden colored fur is soft and fluffy and frames the sweetest face anyone can imagine. Sometimes called the "Golden Tennis Ball," its favorite sleeping position is rolled up in a ball. This position makes it extremely difficult to spot. The round form doesn't resemble any type of animal, and the coloring blends well with the greens, yellows, and browns of the rainforest. In addition, the silky anteater is shy, nocturnal, and lives in the treetops, all of which make sightings rare. It prefers trees with ample liana growth, and when seen from the ground is often asleep in a low-hanging tangle of these thick vines. Limited research suggests that the species is more prevalent than the sparse number of sightings suggest.

As the name implies, it feeds almost exclusively on ants, especially those that live in hollow stems. Using its sharp front claw the silky anteater splits open a hollow stem and laps up the ants with its long, sticky tongue. Those sharp front claws are also a great deterrent to any animal or bird that might have any ideas of harming the furry little animal. The back feet and the prehensile tail both have a powerful grasp, thus freeing up the front feet to slash out at attackers.

One day in February 1994, Juan Ramón, beaming from ear to ear, returned from a tour with three very happy visitors. They had seen a silky anteater. "Get your camera, boss," he said. "It's probably still there."

Away we went, I with camera in hand. About 30 minutes later we came to a tangled loop of lianas hanging from a rainforest tree. And there was the golden tennis ball resembling various shades of dry leaf matter. How Juan spotted it, I will never know. But there it was. I took pictures until I ran out of film. We headed for home, content that the mysterious little animal really existed on Hacienda Barú. But we were in for another surprise.

Arriving at the house we found Pedro who had returned from guiding another tour and was waiting for us with the news that he had spotted a silky anteater, about one kilometer from the one we had just photographed. I was all out of film, but Juan, Pedro, and I went to have a look anyway. Pedro's golden tennis ball was also sleeping in a tangle of lianas. It looked like an exact duplicate of the first one. What an amazing coincidence. After searching and hoping for so long, and then to have both guides sight the charismatic little mammal on the same afternoon.

Since there was a prize involved, we discussed the timing. It turned out that Pedro had sighted his silky anteater at about 1:30 p.m. and Juan Ramón about an hour later. Pedro became the proud owner of *A Guide to the Birds of Costa Rica.*

It was seven years before the golden tennis ball made another appearance, this time on the Night in the Jungle tour. Then we had to wait 11 years before sighting it on a Rainforest Experience hike. Only two years later, on April 3, 2015, I was away from the office and received a call on my mobile. Guides Ronald and Olman had just returned from a Flight of the Toucan zip line tour. "We need a good camera," they told the receptionist excitedly. "We spotted a silky anteater, and it's probably still in the same spot." She called me, explained the situation, and asked permission to use my camera. I readily agreed.

When I returned and saw the photos of the silky anteater, it brought back memories of that first one. Though this one was in a small tree rather than a tangle of vines, it was every bit as cuddly and charismatic as I remembered it to be.

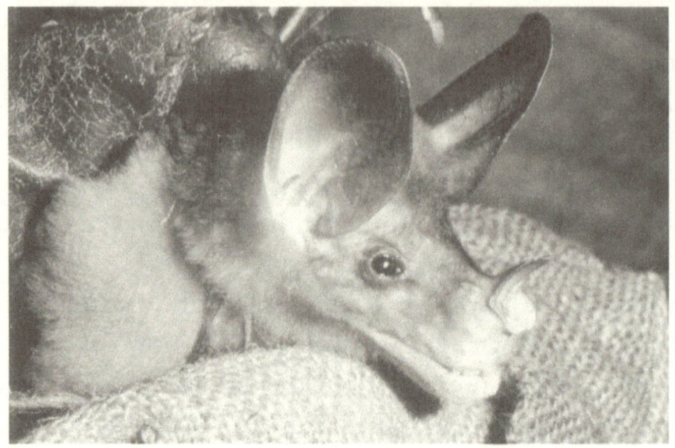

8. What On Earth Is a Chiropterologist?

The Bat Guys

"Would this be a good place to put the net?" I asked Otto. "You said you wanted an open path through the jungle." I stepped into the shallow water of the narrow stream. "I can take one end of the net to the other side and hold it while you secure this end. Then you can join me."

Otto hesitated. "Is there anything in this water that will hurt you," he asked in heavily accented English.

"There are caimans and crocodiles" I replied, "but with all the noise we've been making they're probably all gone by now."

"No, no," he exclaimed, "I mean little animals that live in the water."

"Not that I know of. We walk through this water all the time and have never had any problems."

Finally, he agreed to string the mist net across the stream, but was extremely careful not to get water inside his rubber boots. The natural corridor formed by the stream was a good choice. Over the next hour we captured five different species of bats in the net, including one bulldog bat.

One evening several days earlier I was returning to Hacienda Barú from San Isidro just before dark and noticed three men at the side of the gravel road near the village of Tinamastes extending some very fine nets. I wonder what they're doing, I thought. They look like foreigners.

Two days later a familiar-looking rental 4WD drove up to my house at Hacienda Barú. Three men got out of the car and I recognized them as the men at the side of the road with the nets. The older man introduced himself as Otto. The other two were Fabio and Gunther. They were chiropterologists, he explained, from Erlangen University in Germany.

"What on earth is a chiropterologist?" I asked.

Otto laughed. "Sorry for the big words. It means we study bats. We saw the sign out there at the road that says this is a private nature reserve, and we thought it might be a good place to capture and identify bats. Would that be possible?"

Not only was it possible, we welcomed scientific study of all kinds. They ended up staying with us for a week. That first encounter in 1989 was the beginning of an enduring friendship that lasted many years.

That first night I went with Otto to the stream mentioned earlier, and Fabio and Gunther went to a place I showed them near the house. We used very fine nets called mist nets to capture the bats. The idea was to string the nets across natural corridors where bats normally fly. According to Otto, when flying on a route it knows well, a bat will turn off its radar, or echolocation, and won't detect the mist net. That first night, between the two teams, we captured seven different species. Over the next week we captured another eleven. All the bats were identified and then released. Each new species for the list was photographed before releasing.

The largest bat Otto and I caught that first night was a fishing bat sometimes called the bulldog bat because of its face. It had a wingspan of 80 centimeters. I asked Otto if we were likely to catch any larger bats.

"Probably not," he replied. "There is one carnivorous bat in Costa Rica with a wingspan of about a meter, but it is very rare and has only been seen in one location in Guanacaste. I doubt I'll ever be fortunate enough to see one. It's the largest true bat in the world. When I say, 'true bat' I don't count the flying foxes of Asia." He went on to explain that all the bat species found in North America and Europe eat insects, whereas in Costa Rica we have bats that feed on fruit, nectar, fish, frogs, blood, birds, and other bats. And, of course, we have insect-eating bats.

On the second day we were sitting around after lunch and Otto showed me a cartoon in a biology magazine. It showed two scientific types meeting in the middle of a tropical swamp. One was wearing a swimsuit but was otherwise bare from the waist down. Above the waterline he is completely covered including long sleeves, gloves, and netting over his head. The lower body of the second is covered with chest-high waders but he is shirtless and hatless. "What are you studying," he asks the first? "Malaria," comes the reply. "How about you?" "Bilharzia," says the one with the waders. (Bilharzia, sometimes called schistosomiasis, is a horrible disease caused by a waterborne parasite carried by a snail. It is common in Africa and parts of South America.)

"That's what you were talking about," I laughed, "the other night when you asked if there were any little animals in the water that could hurt you. We're lucky. We don't have any bilharzia in Costa Rica."

"Yeah. That's what I was talking about," he smiled.

Several nights later I sent the three biologists with Gregorio, the forest guard, to spend the night in our jungle camp deep in the rainforest. I thought it would be a great place to find bats. I wasn't able to accompany them but got a first-hand description of the experience when they returned.

Gregorio and Fabio went to a stream where they set up a net. Otto and Gunther stayed at the camp and set two nets. Before they got the second net fully extended, Otto had to return to the first net to free three captured bats. He didn't have time to identify them so he turned them loose in one of the tents thinking he would identify them later. Soon the bats were coming faster than he could free them from the net. Gunther had the same problem with his net. "Take your net down and come and help me," called Otto, "I can't handle all of these bats by myself."

"I can't get the net down," hollered Gunther. "It's too full of bats. I have the same problem as you."

Otto quit taking time to put the bats in the tents and just untangled them from the nets and released them. It was the only way he could possibly empty the net and get it down. With only one net, he and Gunther could probably keep up with the bats. He almost had the net empty when he heard Gunther exclaim, "Oh my God, look at that."

Otto hurriedly untied his net, let it drop to the ground and ran to help Gunther. "Is that what I think it is?" Otto cried out. The net was in tatters with at least five bats in a tangled mess. One bat, much larger than the others was in the middle of the glob eating one of the smaller ones. "It is. It really is a *Vampyrum spectrum*. I didn't think I would ever see one. Hurry, let's get him out of the net. Stay clear of those teeth."

It took them 20 minutes to free the bats. Otto held the false vampire bat by the wings while Gunther photographed it. They decided to keep it in one of the tents overnight so they could examine it more closely in the morning and get more photos.

"Sorry about your net," said Otto. "It's beyond repair. That monster tore it to shreds. You'll have to get a new one."

Gunther looked at him with a devilish grin. "We are going to have some explaining to do," he replied. "That was Fabio's net. He took mine by mistake, so I used his. He's not going to be happy about this."

"Well, maybe we can all go together and share the expense of a new one. After all, this is quite an accomplishment for the whole team. We can publish a paper in one of the journals. Just think.

This one bat made the cost of the whole trip to Costa Rica worth it. We're not going to worry about losing a net."

Otto and Gunther didn't put any nets back up. Instead, they went about identifying the bats they were holding in two of the tents. They left the tent holding the false vampire bat closed so as not to risk losing it. The task would be much easier in the daylight.

Shortly after they finished Fabio and Gregorio returned. "How did it go?" asked Otto.

"Not too bad," replied Fabio. "I got another new species for the list. How about you guys. You catch anything interesting?"

"Well," said Otto, "this is one of those good-news, bad-news stories." Fabio just looked puzzled but didn't speak. "We got four new species."

"That's great," said Fabio, still dubious.

"Including a *Vampyrum spectrum*," interjected Gunther.

Fabio stared in disbelief. "You're putting me on, right? You didn't really catch one?"

"Yes, we really did," said Otto solemnly. "The bad news is that we caught it with your net." He pointed his flashlight at the tattered mass of filaments on the ground by one of the tents.

Fabio broke into a broad grin. "This is unbelievable. Tell me all about it. I'm dying to hear."

"We can do better than that," smiled Otto. "We still have him. He's in that tent over there."

The three chiropterologists slept outside that night, next to the tent with the world's largest true bat. They said they didn't sleep much. They rested and dozed listening to the sound of crunching bones as it devoured the four other bats in the tent.

Dr. Otto Helversen returned to Hacienda Barú many times over the years, as did several of his graduate students. To date they and other researchers have identified 38 species of bats on what is now the Hacienda Barú National Wildlife Refuge. We call them "The Bat Guys."

9. Bird Watching Fever: Ecotourism At Its Best

The Dance of the Red-Capped Manakin

For the uninitiated, a bird watcher's enthusiasm for hearing, sighting and studying our feathered friends is hard to fathom. Why would anyone rise with the sun, walk for hours while toting binoculars, spotting scope, tripod, field guide, checklist and notebook just to observe birds?

We all see birds every day, so what's the big deal? Hard-core bird watchers have even been the brunt of jokes and cartoons. One of my favorites is a Gary Larson Far Side Cartoon with a view through a pair of binoculars, showing a large nasty looking bird sitting in an equally large nest, staring evilly at the observer…the person looking through the binoculars. Dangling from the edge of the nest are several pairs of binoculars, a birder's hat and a tote bag with a *Field Guide to the Birds of America*. No caption was necessary.

Seriously though, bird watching is big business, and Costa Rica is a hot spot for birders. To begin, it is located in the tropics, a warm hospitable climate, midway between the North and South American continents. Migrants from both may find a warm haven during their respective winters. It also has coastline on two oceans. In altitude Costa Rica's terrain ranges from sea level to nearly 4,000 meters, well above timberline. In the tropics, a slight change in altitude brings a corresponding change in climate. There are more than 20 micro-climates in Costa Rica, and *A Guide to the Birds of Costa Rica* by Stiles and Skutch, list 20 distinct habitats found in the country. Because of this diversity of environmental conditions and factors, Costa Rica is host to over 830 species of birds, about the same as the United States and Canada combined.

As any ecotourism establishment well knows, bird watching enthusiasts come from every part of the world. I remember an afternoon in January of 2004 when all six of Hacienda Barú's cabins were rented to bird watchers from as many different countries. Prior to setting out for their late afternoon birding hikes, everyone had congregated in the garden to compare notes and checklists. Spotting scopes, field guides, cameras and binoculars surrounded the tables where the aficionados were gathered. The enthusiasm was contagious as people from Costa Rica, the United States, Canada, Germany, Spain, and England talked excitedly about the one thing they all had in common, an intense interest in birds.

I have long admired British bird watchers. This is partly because their approach to birding is so methodical, but also because my first birding experience with experts was with a British couple, Roger and Sharon. The experience took place in the late 1980s when we had just begun to cater to ecotourism at Hacienda Barú. The couple stopped by one afternoon and scheduled a hiking tour called "The Rainforest Experience" for the next day. When I told them that they would see many more birds in the lowlands on the "Mangrove Walk," Roger agreed, but said the birds would probably be ones they had already logged during their two weeks in Costa Rica. What they wanted to see were the rainforest birds, fewer species in a habitat with limited visibility and unlimited hiding places. We scheduled the "Rainforest Experience" for the next morning, before daybreak. Little did I suspect that within 24 hours I would be afflicted with bird-watching fever; that I would return from the hike a full-fledged addict.

We made it to the rainforest campsite about the same time as the first crimson rays of sunlight filtered through the rainforest canopy. Over a breakfast of coffee and sweet rolls we listened to a cacophony of jungle wake-up sounds, hundreds of creatures struggling to make themselves heard. No small portion of the calls came from birds. Roger would cock a thoughtful ear to one side for a moment or two and pronounce the name of the bird that had made the sound.

The most interesting sighting at the jungle camp was a pair of mealy parrots, large impressive birds that squawked outrageously for the duration of breakfast. It was a first for me, but Roger and Sharon had already logged it elsewhere. Preparing to set out through the forest, Roger described the type of habitat they were looking for, areas with thick understory and vine growth. Antbirds and manakins were what they really hoped to see. I pretended to know what they were talking about.

Roger's highly trained eyes and ears picked up many birds that I missed. I saw all the toucans, motmots and big red-headed woodpeckers, but those didn't seem to impress my clients. Instead, they seemed more excited about some of the little birds that I tended to dump into the single category of "little brown jobs" or "LBJs." But, as I listened to Roger and Sharon, their enthusiasm captivated me. My education was only beginning.

We passed through an area where the understory seemed denser and the vine growth thicker. I heard a loud snapping sound but couldn't imagine what might be making it. As we got closer, I noticed that the "snap" was preceded by a "buzz." Ever more curious, I carefully scanned the surrounding foliage for the source of this strange disturbance. I spotted the manakins almost at the same time as Roger.

"Unbelievable! Absolutely amazing! I have never seen this," he whispered. "This must be the mating dance of *Pipra mentalis*, the red-capped manakin. Brilliant!"

The setting was as perfect as a carefully set stage arranged precisely for the actors that were performing in this pristine theater. At center stage sat the dull, light-greenish, rather nondescript female, preening herself while four suitors took turns competing for her favor. Her throne was a small twig. Four longer branches, each coming from a different direction, extended inward toward the princess, with one aspiring beau on each.

One by one each of the males danced the length of his branch, starting on the far end with his bright red head bobbing, his vibrant orange thighs and yellow legs in a flurry of movement, carrying his coal black body the length of his branch to within a few centimeters of his lady love. At this point he emitted a loud "buzz," leaped into the air and made the loud "snap" that had attracted me to the scene. The "snap" was so quick that it was impossible to discern how he produced it. I remember thinking that it might be by clapping the wings together, but then discarded that possibility thinking that the force necessary to make such a loud "snap" would certainly break wing bones.

The fervent suitor repeated his dance in an outward direction, finishing with another buzz and snap. Then he waited quietly, the perfect gentleman, while his rivals made their bids. During the entire performance, the object of all this attention, the lady in waiting, appeared not to realize that she was the big attraction, and totally ignored all four aspirants. After each proud performer had received two or three opportunities, the seemingly unimpressed female stopped preening herself, sat up, looked around and flew away. The four males looked at each other, totally dumbfounded, jumped into the air and flew after her. I couldn't help but chuckle.

Roger explained that the stage for this incredible mating game is called a lek. *A Guide to the Birds of Costa Rica* confirmed that the loud "snap" was, indeed, produced by slapping the wings together. Each species of manakin has a slightly different version of the dance. Though four other species of manakin are found on Hacienda Barú—blue-crowned manakin, white-ruffed manakin, orange-collared manakin and thrushlike manakin—even Roger admitted he couldn't tell the females apart. The males, however, are quite distinct. With the exception of the thrushlike manakin, all of these forest denizens have leks and ritual dances, and make a high-pitched peep each time they jump up into the air from their perch. In Spanish their common names tend to suggest jumping or dancing, *saltarin* or *bailarin.* They are all small and beautiful, but none quite so charming as my red-headed favorite.

Though we sighted several more birds, the walk home was uneventful in comparison to the spectacles we had already witnessed. Though I was supposed to be the guide, in all honesty, I did little more than keep us from getting lost. Roger and Sharon were the experts, and they graciously shared their knowledge as well as their enthusiasm. Never again would I wonder at the fervency of the typical bird watcher. That day was my initiation into a new world.

Bird watching means more to me than the excitement and satisfaction that comes from the pursuit and identification of avian species. I firmly believe that the conservation and regeneration of the rainforest is of utmost importance for the future of our planet. I also believe that the only way conservation will work is to make it profitable. There simply aren't enough philanthropists in the world with enough money to save these critical natural environments. If however, we can find ways to make a living from intact rainforests, without harming them, we will have a built-in incentive to protect them. Responsible ecotourism is a proven method for doing this.

Without ecotourism we could not afford to protect Hacienda Barú National Wildlife Refuge and maintain it in its natural state. Bird watching is a very special kind of ecotourism that accounts for a large percentage of our visitors. Give it a try. But be careful, it's addictive.

10. The Spider Monkey and the Garlic Tree

One of Nature's Many Puzzles

Fluffy yellow flowers carpeted the trail. How beautiful. Then the odor overwhelmed my nostrils. Garlic! "Oh my God," I exclaimed turning to my friend, Juan Ramón. "Are these flowers from the ajo tree?"

Juan laughed. "There it is, right over there," pointing to a tall, thick, straight tree about 20 meters off the trail. "Haven't you ever seen the flowers before? I know you love the tree."

The tree is not only enormous, but also tall, thick, and straight. The wood is strong and very resistant to water. Ranchers sought them out, felled them and used the wood to make boards for corrals. It was also one of the preferred woods used for railroad ties when Costa Rica's railroads were being built. Other uses include structural supports for bridges and buildings. In the last century so many of them were cut that very few are left.

When you hear of a species in danger of extinction, an image of a charismatic mammal, like a panda or a tiger, usually comes to mind, not a tree. Yet one of the biggest, most impressive trees I have ever known has been listed as vulnerable on the IUCN Red List since 1998, *Caryocar costarricense*. As the name implies the tree is found mostly in Costa Rica, though there are a few of them in Panama and possibly in Colombia. As the species is not found in any country where English is spoken, I have never heard of a name for it in that language. The Spanish common name is *ajo* (pronounced "AH-ho") or *ajillo* meaning garlic.

As I write, Hacienda Barú still has eight individuals that we know of, and there may be a few more in remote parts of the reserve that we haven't found yet. I have known of the plight of the ajo tree for many years and had hoped that our trees could serve as a supply of seeds to help replenish the species. I soon found out that all of our ajo trees were mature, and there were no seedlings or medium-sized trees to be found. The mature trees produced fruits with seeds, but those seeds apparently didn't germinate.

One self-proclaimed expert in plant nurseries told us that the seeds will germinate, but only if they have been cut from the top of the tree and have never fallen to the ground. This sounded a bit far-fetched, as the seeds surely germinate in nature, under certain conditions, and that germination is likely to take place on the ground, not in the top of the tree. But he was the expert, so one our tree climbing guides managed to ascend more than 40 meters into the top of an enormous ajo tree and cut a bunch of seeds. The "expert" took them to his nursery to work his magic on them, but only two seeds out of several hundred germinated, and the seedlings died before they reached 20 centimeters in height. So much for the tree-top germination theory.

In the mid-1990s, when I was still guiding people through the rainforest, I showed an ajo tree to a group of visitors and explained that extinction was a possibility for the tree. The lack of germination almost guaranteed that the species would eventually become extinct on Hacienda Barú. One of the ladies in the group spoke up. "A few days ago, we were in Carara National Park and saw an ajo tree," she explained. "Our guide told us about the reproduction problem and mentioned that the seeds won't germinate in a forest unless spider monkeys are present. Are there any spider monkeys here?" I told her that the white-faced capuchin was our only monkey species, and questioned her further about the incident in Carrara, but the guide had only made the one brief comment.

This was very interesting. Puzzle pieces were starting to shift around, coming closer to fitting together. Many years ago, spider monkeys, howlers, and capuchins all abounded at Hacienda Barú, but around 1947 an epidemic of yellow fever swept through this area and wiped out all of the first two species, leaving only the white-faced capuchins. There hadn't been any spider monkeys on Hacienda Barú for 50 years, and all of our ajo trees were more than 50 years old. Then in 1997 a lone male spider monkey appeared, then a female, then a couple more. New secondary forest was beginning to appear along the coastal ridge, and the Path of the Tapir Biological Corridor was starting to take shape and connect isolated forest patches together. The spider monkeys were the first indication that the corridor was working. Later the howlers came too.

When I decided to write this essay, I mentioned to Hacienda Barù guide Deiber Segura that I needed a photo of one of our more impressive ajo trees. I knew that he was going to be near it that day. I told him a little bit about the article and the tree. "Did you know that there is young ajo tree along the ridge on the left side of the toucan trail?" he asked.

"How do you know it's an ajo?" I asked, trying to keep the excitement out of my voice.

"Because the leaves have a sawtooth edge," he replied, "and the stems are red."

I went to see the tree the next day with Juan Ramòn, who was the only guide old enough to know what a seedling should look like. It took us a while to find it, but once he spotted it Juan knew immediately. "It's been so many years since I have seen one of these," he said. "And it's the first on Hacienda Barú."

The tree was about three meters tall, and still quite thin. It looked healthy and, like Deiber said, had sawtooth edges on the leaves and red stems. Later Deiber told me he had seen another young tree almost a kilometer from the first.

I found this quite interesting and the puzzle pieces moved a little closer together. The first spider monkey returned 21 years ago, and now we are starting to see young trees. The evidence wouldn't hold up to scientific scrutiny, but it was a strong indication that spider monkeys might have something to do with ajo reproductivity. What we need is to find out if the ajos reproduce in a forest that has both mature ajo trees and spider monkeys, a forest like the one on the Osa Peninsula.

Searching the internet for more information I came across a study done by Silvia Solis, Jorge Lobo, and Maryori Grimaldo on the Osa and published in the *Academic Journal Biología Tropical.* Not only did the study confirm that the ajo trees on the Osa reproduce regularly, but the researchers observed female spider monkeys with young eating the pulp off of the fruit and dropping the seeds on the ground. Scarlet macaws were also seen eating the fruits and, presumably the seeds as well. This was especially interesting since the big red birds have been locally extinct since the early 1960s and just started visiting Hacienda Barú again in 2014.

One particularly interesting bit of information that came out of the Osa study is that medium-sized trees produce the most fruit. That would be those with a diameter of between 80 and 190 centimeters. The researchers found that trees outside of that range produced little to no fruit. The diameter of the big tree in the photo with the two men is 106 centimeters, well within the optimum size range for fruit production.

This is good news, because that tree, and others on Hacienda Barú, all of which are a similar size, will be in that optimum range for many years to come. During that time, we expect populations of spider monkeys and scarlet macaws to continue to increase and, if in fact they do play a part in the germination of ajo seeds, their presence will likely promote seedling production. Maybe the future of the ajo tree on Hacienda Barú isn't as bleak as I had once imagined.

This would make a wonderful project for a graduate student who would like to spend a few years studying *Caryocar costarricense.* We are far from having proof, but we do have enough evidence to support the hypothesis that the presence of spider monkeys and/or scarlet macaws is essential to the reproduction of the tree, but there is much work to be done before any level of certainty can be obtained. The big reward at the end would be if we could learn enough to save a species from extinction.

It is one of Mother Nature's many fascinating puzzles.

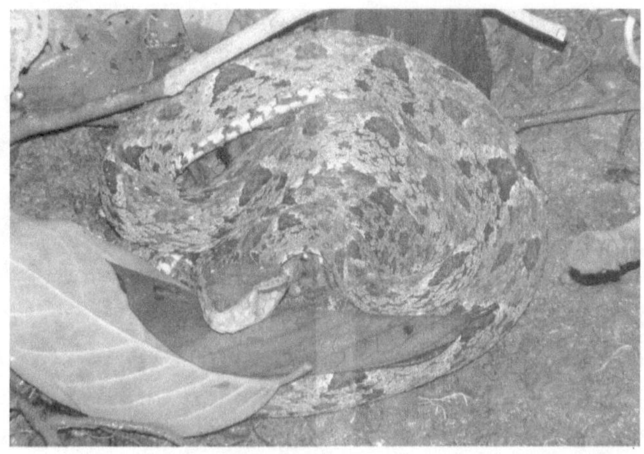

11. If You're Unfortunate Enough to Get a Snake Bite

...Do It In Costa Rica

An afterthought at the end of the email from my brother Rex read, "..and our dog Maggie has been bitten on the muzzle by a rattlesnake."

Knowing that the only poisonous snakes in that part of Colorado are prairie rattlers, I figured that Maggie would survive. These small pit vipers are not nearly as dangerous as some of the other species, but can still kill. I wrote Rex to inquire about the bite. In the correspondence that ensued it came to light that Maggie had been vaccinated against rattlesnake bite, the vaccine cost about $25, and apparently gave her a partial immunity. Other than severe swelling in her face for about 24 hours, Maggie recovered just fine. It wasn't necessary to take her to a veterinarian. Apparently, there is no vaccine for people.

One further bit of information inspired me to delve into the subject of snakebites and write this article. In the two decades since Rex and his wife LaVonne have lived in their present home, three of their neighbors have been snake bitten, and the cost of treatment ranged from $70,000 to $100,000. I knew that snakebite serum in Costa Rica is very good and very inexpensive, and I wanted to know why it costs so much in Colorado. What I found is nothing short of astonishing.

Instituto Clodomiro Picado

Dr. Clodomiro Picado Twight (1887–1944) was a medical doctor who worked at the San Juan de Dios hospital in San Jose. He is best known for his extensive research into venomous snakes in Costa Rica and the treatment of snakebite. The institute that honors this distinguished physician was founded in 1970. Its creation was a joint effort of the Ministry of Health and the University of Costa Rica.

I first visited the Instituto Clodomiro Picado in 1972. It was open to the public on Saturdays, and the demonstration of milking of the large venomous serpents was fascinating. One of the scientists explained how they produce the antidote for treatment of snakebite. Small doses of venom are injected into horses, and the dose is gradually increased as the horse's immunity builds. Eventually the horse becomes totally immune to the bites of the most common pit vipers found in Costa Rica. Blood is drawn from the horses, and the cells separated from the serum in a centrifuge. The antibody-laden serum is the base for the anti-venom, called antivenin. He explained that there was a tremendous demand for the serum especially in the rural areas where low-income farm workers make their living working in fields and pastures frequented by poisonous snakes. It was the mission of the institute to produce enough high-quality serum to treat anyone who needs it. We went away from the demonstration full of admiration and awe.

The Instituto Clodomiro Picado has grown and evolved into a prestigious institution that all Costa Ricans can be proud of. In the year 2014 the facility produced 90,000 vials of antivenin, about half of which was used in Costa Rica. Additionally, their products are sold throughout Central and South America and even in Africa.

Costa Rican hospitals treat an average of 600 snakebites annually, mostly terciopelo. Despite the large number of bites there are only one or two deaths, testimony to the excellence of the antidote.

Most in demand is a product called Suero Antiofidico Polivalente. This is effective in the treatment of bites from most of the pit vipers of Central America. It is elaborated with the venom of the three most common pit vipers in Costa Rica: terciopelo (*Bothrops asper*), bushmaster (*Lachesis stenophrys*), and rattlesnake (*Crotalus simus*). Polivalent serum is available in liquid form with a shelf life of three years when refrigerated, and in a dry form that is mixed with sterile solution shortly before using. The dry form has a shelf life of three years without refrigeration. The institute manufactures a similar product for veterinary use, and another that is specific for coral snakes. They have even elaborated a serum that is effective in the treatment of the bites of some of Africa's venomous serpents. This last product is exported mostly to Nigeria. The Polivalent Serum is sold in Costa Rica for $19 to $20 per vial.

A friend of mine was bitten by a terciopelo in 2002 and was treated with the Polivalent product, of which she received a total of 10 vials. She was hospitalized at the Max Terán Valls hospital in Quepos for three weeks and her total bill for everything amounted to $10,500. The cost of the antivenin was insignificant. She mentioned that the hospital was very proficient in administering the treatment, and that during the three weeks she was there, four other patients were admitted with snakebites.

Snakebite Treatment in the United States

I was curious as to why the treatment of the rattlesnake bites suffered by my brother's neighbors was so costly, so I did a little digging. To date, the FDA has approved only one antidote for the treatment of snakebites in the US, a product called CroFab. It is produced by a British Company called BTG Plc which, according to their web site, was named as one of Forbes.com's most innovative growth companies. CroFab is similar to Instituto Clodomiro Picado's Polivalent Serum in that both are produced by immunizing mammals to the bites of several different snakes and elaborating the antidote from the antibody-laden blood serum of those mammals.

But there are some differences, the most notable being that BTG uses sheep rather than horses. Another difference: Instituto Clodomiro Picado has their serpentarium, horse farm, and laboratory all in one place where they produce their snakebite antidotes. BTG has the venomous snakes in the US, the sheep in Australia, and the laboratory in England (source: *Bloomburg Business).* That's right. You didn't read it wrong. They milk the venom from four different kinds of venomous snakes in Utah and ship it to Wales for processing. From there it is shipped to Australia where it is used to immunize sheep against snake bite. Blood serum from the sheep is then shipped back to Wales where CroFab is manufactured and shipped to hospitals in the United States. The list price is $2,700 per vial, but the discount price to hospitals is about $2,200 per vial.

CroFab is apparently a very good product as the ratio between snakebites and deaths is about the same in the US, as in Costa Rica. Estimates vary, but one source tells us that in the US there are approximately 5,000 bites per year and only 10 to 15 deaths. There may be some competition on the horizon as a Mexican firm called Rare Disease Therapeutics is producing a product derived from horse serum that sells in Mexico for about $100 per vial. This product will supposedly be available to US hospitals soon.

But there is more than just the cost of the antidote. Reported in the *Charlotte Observer* and *Time Magazine* was the story of Eric Ferguson's snakebite treatment. After being bitten by a snake, Mr. Ferguson reported to the emergency room of a hospital in North Carolina. The bite wasn't too serious, and he needed only four vials of antidote. He was released after 18 hours. His bill was $89,277. He had been charged over $20,000 per vial for the antidote.

Mr. Ferguson did his homework, found out how much the antidote costs, and demanded to know why he was charged so much for a product that cost the hospital much less. A hospital spokesperson explained that the difference in price was due to the cost of storing, preparing, and administering the antidote. Also the cost of dealing with Medicare, Medicaid, and insurance companies was mentioned.

It is beginning to look like even if the Mexican product does become available in the US for $100 per vial it probably won't affect the price to the patient significantly. Most of the costs are hospital charges and not directly related to the value of the antidote. Eric Ferguson received a very minor bite, but there are other more severe cases that require a great deal more serum. Eleven-year-old Benjamin Smith received a massive bite from an eastern diamond-back rattlesnake and needed 80 vials of CroFab. His hospital bill was $1,600,000.

Several promising new techniques are being researched, but a product for treatment in humans is still in the distant future. One possibility now under study by the Instituto Clodomiro Picado is the use of chickens rather than horses to produce the antibodies that are the basis of the antivenin. Antibodies from the chicken's blood are passed on to the yolks of their eggs. From there they can be extracted for use in the manufacture of antidotes.

According to *Chemistry World* a number of researchers are investigating a peptide found in the blood of opossums. It has long been known that opossums are amazingly resistant to snakebite. The peptide responsible for the immunity can be produced inexpensively by genetically engineered bacteria. Though still a long way from the production of a usable product, the peptide has been tested in laboratory mice with very encouraging results.

According the World Health Organization, 1.8 million snakebites occur annually worldwide resulting in 94,000 deaths. Most of the deaths are people from third-world countries who don't have access to treatment. Please don't become one of those statistics. When in snake country, use adequate foot wear and watch where you step. But if you do have the misfortune to become one of the 1.8 million, I recommend that you do it in Costa Rica.

12. Ocelots, Tayras and More Agoutis

A Fun New Way of Monitoring Wildlife

"Oh my! Absolutely incredible," I blurted out to no one in particular. A full-grown puma was walking straight toward me on the computer screen. The two receptionists who were in office heard my exclamation and hurried over to see what I was so excited about. I restarted the 15-second video so everyone could see. An adult puma came walking straight at the camera, its eyes glowing from the infrared illumination. It looked like it was stalking the camera. At the last moment the large carnivore veered to the right and walked out of the field of view.

It was a very exciting day. The presence of large carnivores in the Hacienda Barú National Wildlife Refuge was a clear indication of a healthy ecosystem. Since that first video in December 2012, we have captured around 200 pumas at nine locations within the refuge.

Trail cameras, also called camera traps, have been around for at least 35 years. The cost of early trail cameras was so high that few researchers could afford them. The digital revolution brought revolutionary change to the cameras which quickly improved, and prices got lower. By 2010 prices ranged from several thousand dollars to less than $200. Camera traps were becoming affordable to the average nature lover. The cameras I use today are made in Asia and cost $40 each. They are triggered by an infrared sensor that detects warm-blooded animals. The photos are stored on an inexpensive memory chip which holds thousands of photos or videos.

My first camera trap was a gift from my daughter. It was the beginning of a great adventure. Programing it was a real challenge for me, but once I had studied the instructions and gotten into the process, it went well. There were settings for date and time, choice of video or photos, and resolution of the images. If you choose photos, you can program the camera to take one at a time or three in succession. I tried it out in the house taking photos of the dogs. It worked perfectly. I was ready to put it out in the rainforest. One of the guides suggested a location where there is a short bridge across a jungle stream and lots of animal tracks. We tied the camera to a tree at one end of the bridge.

I waited a week before retrieving the memory chip. With great anticipation I put the chip in my computer. The first photo appeared on the screen, me backing away from the camera after turning it on. The next one was a coati approaching the bridge. I was elated. Until that moment I had been doubtful, but now I knew that it really did take pictures of wild animals. On that first chip there were photos of four different species in addition to the coati: skunk, paca, four collared peccaries, and about a dozen agoutis. I couldn't have been happier. I called my daughter and thanked her again for the gift.

Several months later I brought back a chip that had about 50 jpg files on it, but for some reason my computer wouldn't open them. I tried everything I could think of, but the files wouldn't open. I finally decided that it was a lost cause, and almost deleted the contents of the chip. At the last moment I tried it with a different computer. To my delight it worked. And to my astonishment the very first image to pop up on the monitor was an ocelot.

I just sat there staring at the screen. After recovering from the initial shock, I called everyone over to see. Finally, I got down to studying the photo. I decided that it was a female. Although she was walking at an angle toward the camera, and her rear end wasn't visible, her udder was visible and was quite full. I surmised that she was either in late pregnancy or nursing a cub or cubs. There was a black object in her mouth, but the image was a little fuzzy. Several Hacienda Baru guides studied the image, and we determined that it was a spiny rat, the ocelot's favored prey. Since that time my cameras have captured photos and videos of at least 250 more ocelots at about 15 different locations within the Hacienda Barú National Wildlife Refuge.

Other than being fun, trail cameras are often used in scientific research. For example, researchers working at Hacienda Barú have used them to study animal behavior, diversity and frequency of mammal species, and monitoring locations where animals cross the highway through tunnels and over bridges. The Asociación Amigos de la Naturaleza del Pacifico Central y Sur (ASANA) has used them to determine the level of diversity of mammal species in the Path of the Tapir Biological Corridor (PTBC).

When highway #34, which passes right through Hacienda Barú National Wildlife Refuge, was in the planning stages we worked with the environmental impact study team to devise ways to mitigate the roadkill. We asked them to build tunnels under the road and suspension bridges over the top. The biologists in charge were very cooperative. We ended up with 21 tunnels and four bridges in a two-kilometer stretch. We knew that animals had started using the tunnels almost immediately because we could see their tracks, but no one had actually seen or photographed one. We assured highway department that the project had achieved its goal, but that we had no proof to back up our words.

I moved one of the six camera traps to a tunnel and I set the camera to take 15-second videos. Usually, I check the cameras once a month, but I was so anxious to see if that camera had captured any evidence that after five days I retrieved the memory chip. The results blew me away. There was a total of 33 videos of four different species walking through the tunnel: peccaries, pacas, coatis, and, of course, agoutis. The most impressive was a group of about twenty peccaries walking toward and past the camera. I posted it on the Hacienda Barú Facebook page. The video went viral.

Messages and emails from interested individuals, government officials, biologists, and numerous environmental groups started pouring in. Some came from faraway places like Brazil and Argentina. Costa Rican environmental groups started pressuring the highway department to build animal crossings at their locations and even approached the national legislature about creating a law requiring that tunnels be included in all new highway construction that passes through a national park or wildlife refuge. The people from the highway department who had helped get the project approved were elated. Now they had proof that the funds had been well spent.

The more I worked with trail cameras the more I learned. That first Moultrie brand camera took excellent quality photos and videos. The main drawback was the visible flash for nocturnal photos. Animals didn't like it. Whenever I mounted the camera in a new location it would capture lots of photos or videos the first week. As time went by there were fewer nocturnal images. I reasoned that the animals quickly learned that every time they walked past that location a bright light would flash in their eyes, so they changed their habitual route to avoid it. Later cameras started appearing on the market with infrared illumination. These were better, but there was still a red glow from the infrared LED bulbs. Today many brands offer cameras that have no visible light at night.

Learning where to place the cameras so they will capture the most and best quality images comes with experience. Wildlife often follow walking trails used by people. One important trick is to aim the camera down the trail rather than across it. Once the infrared sensor triggers the shutter there is a delay before it starts recording images. An animal walking past a camera can trigger it and be beyond the field of view by the time the photo is snapped. An animal walking toward or away from the camera will still be visible. Though they are water resistant, the useful life of an unprotected trail camera in a rainforest is only about eight months. If you put a cover over it, the life is extended to about two years. We fashion covers out of plastic gallon jugs for protection, an easy and inexpensive solution. Even with the cover I bring my cameras in after a couple of months of use and give them several days of hot sun to eliminate any internal humidity.

Our best camera location on Hacienda Barú is where five trails meet. It is where the camera is most likely to capture a rare photo or video of a tayra, and usually one or two ocelots per month. Ocelots are lots of fun because every spot pattern is different, making it relatively easy to identify individuals.

This year from March 10 to April 10 there were two, an adult female and a grown cub traveling separately. Probably the cub had been weaned and was on its own but still sharing the same territory with its mother. Over the next month we saw the same two, but additionally there were three adult males, including the one that normally occupied that territory. One of these males was so large that at first glance I thought it was a jaguar. On June 1 at 7:20 p.m. the female stopped right in front of the camera, licked her paw, and continued onward. Thirty-four minutes later, the big male stopped in the same spot, sniffed the ground, and continued on. Obviously, he was following her.

I discussed this with a biologist friend who has several cameras about 50 kilometers south of here. His cameras had captured photos of three different males in April. We decided that the mating season had begun, the female was coming into estrus, and the big male was following and biding his time until she was ready to mate. He had probably ousted the other two males from the territory. Due to this experience, we now know that the mating season for ocelots in this part of Costa Rica includes April, May and June. I am anxious to see the offspring of that enormous male.

At the top of my wish list is the tapir. The Path of the Tapir Biological Corridor (PTBC) project was initiated in 1990. The last tapir of record in the area was shot by a hunter in 1957. There are tapirs in the Los Santos Forest Reserve at the north end of the corridor and in the Osa Peninsula to the south. The dream of the project is to restore enough natural habitat to the area so that these large mammals will return. If a tapir were to appear on one of my cameras or those of anyone else within the corridor, it would be a clear indication that the dream is well on its way to realization. There are three other mammals on my wish list: jaguar, white-lipped peccary, and brocket deer, all of which are very scarce or locally extinct.

During June 10 to July 10, 2016, the Local Council for the PTBC placed cameras at certain locations along the length of the corridor with the goal of determining the presence and distribution of mammals. Also included were photos and videos of cameras belonging to individuals who are cooperating with the project. My best camera was included. Before I sent the chip to the project coordinator, I copied it to my computer. There were 367 videos, 300 of which were agoutis. Fourteen other species appeared on the chip: four birds (great curassow, great tinamou, blue-crowned motmot, and a white-tipped dove), and 12 mammals, including collared peccary, tayra, skunk, tamandua (anteater), squirrel, spiny rat, white-tailed deer, opossum, raccoon, paca, coati, and ocelot. Of the ocelots, four videos included two different males, the female, and the grown cub.

Over the years my cameras have captured more photos of agoutis than any other species. Every time I stick a chip in a computer to view the videos, I am anxious to see what appears. Nevertheless, I am aware that in between the exciting species there will be agoutis, agoutis and more agoutis.

Trail photos: A rare sighting of a tarya (similar to a big weasel) and an agouti.

13. The Bush Dog

A New Mammal in Costa Rica

Rarely is a new mammal discovered anywhere on the planet. Insects, yes. In fact so many new insect species are being discovered every day that it is not even newsworthy. Species of reptiles and amphibians new to Costa Rica are found at least once a year, birds occasionally, but 2016 was the first time a new mammal species has been confirmed in Costa Rica during the nearly five decades that I have lived here.

I am referring to the bush dog (*Speothos venaticus*), a member of the canidae family sometimes called a savannah dog or vinegar dog. A video of a pair of them was captured on a trail camera by biologist Jan Shipper in the Talamanca region at 1,493 meters above sea level. Prior to that time there had been several unconfirmed sightings in the Osa Peninsula, but no photos. It was believed that Shipper's video was taken at the highest elevation where bush dog presence had been detected anywhere throughout its range. A couple of years later Dr. Mike Mooring, of the Talamanca Large Mammal Survey, captured photos of several groups of these canids at an elevation of 2,086 meters, considerably higher than Shipper's video. Prior to Jan Shipper's video, the species was not known to exist north of Panama.

It is possible that bush dogs have been present in Costa Rica for a long time but were never detected. However, Shipper had been working with trail cameras in the Talamanca region for 12 years prior to that first video without detecting a single one. It seems likely that they are migrating into remote parts of Costa Rica from Panama and adapting to higher elevations where there is less human encroachment. Though they are known to exist in all the countries between Panama and northern Argentina, and in many different habitats including mangrove, swamp forests, savannahs, secondary forest and primary forest, sightings are rare, and little is known about these fascinating carnivores.

Fossils of *Speothos venaticus* were discovered and the species named many years ago by Peter Wilhelm Lund. He found them in caves located in Brazil. The species was thought to be extinct. Today we know that not only are they not extinct, but that they are quite

Costa Rica trail cam photo of three bush dogs. (Mike Mooring)

widespread. Nevertheless, sightings are few, and research is almost non-existent, mainly due to the difficulty of locating and studying the species. Much of the available knowledge about their habits is anecdotal or comes from the observation zoo animals whose behavior can be markedly different from wild animals.

Bush dogs exude a strong odor similar to vinegar which gives rise to the nickname "vinegar dog." The odor is apparently very distinctive, as the same nickname occurs in three languages throughout the species' range. They have been described as about the size of a terrier. Wild animals weigh around 8 kilos and stand about 30 cm at the shoulder. The body is long and the tail short. The thick coat is reddish brown to black, being lighter on the head and back, and black on the tail. Bush dogs have webbed feet and are excellent swimmers.

Most hunting is done in packs, allowing them to subdue species larger than themselves, including pacas, agoutis, armadillos and peccaries. There is even one report of a pack of bush dogs harassing a tapir. Six to ten individuals consisting of an alpha pair and a number of apparent family members comprise the pack. Only the alpha pair mate and reproduce, and the entire pack cares for the offspring. According to studies of zoo animals, the alpha female uses hormones to prevent other females from coming into estrus. In captivity litters of as many as 10 pups have been recorded, but in the wild the number is usually four or five. Like domestic dogs, newborns are blind and helpless. Hollow logs and burrows made by other animals, such as armadillos and pacas, serve as dens.

Hopefully the bush dogs will find a niche in Costa Rica, remain here permanently and expand their range. They have demonstrated a great deal of adaptability by living in many different habitats and climates at different elevations. It is even conceivable that they could migrate as far as the Path of the Tapir Biological Corridor in the coastal region south of Quepos. There is a continuous forest corridor from the Amistad International Park near the Panamanian border all the way to the los Santos Forest Reserve northeast of

Quepos. It would mean migrating into even higher elevations than previously documented, but adaptability is one of *Speothos venaticus'* strong points.

As with other mammals, human encroachment on their habitat is a major threat. Another is reduction of prey animals populations due to illegal hunting. Hunting of any species is illegal throughout Costa Rica, but enforcement is lax. Unless this changes, lack of prey could be a serious threat to bush dogs. A third threat comes from another canid species, domestic dogs. Bush dogs have been known to contract diseases such as distemper, parvovirus, and rabies. Fortunately, they don't seem to have an affinity for human settlements, and this threat may not become a major problem.

14. King of the Jungle

Big, Beautiful Bird, Stinky Feet

Riding along the edge of the forest on horseback I caught a whiff of something rotten. A moment later a big, beautiful, black and white bird with a multicolored head came into view. It was pecking away at the stinking carcass of a dead opossum. A bunch of black vultures were hopping around nearby, mostly just watching, but occasionally darting forward and snatching a morsel of the decomposing flesh before quickly withdrawing. All kept their distance from the magnificent creature which calmly ate its fill, paying little attention to the others.

"What is that beautiful bird?" I asked Orlando. "The zopilotes sure give it plenty of space."

"Of course they do," replied Orlando, a slight smile on his face. "He's the king. We call him El Rey de Zopilotes (King of the vultures)."

"I never knew a vulture could be so beautiful," I blurted out in amazement. "I can see why you call him the king."

Throughout history the king vulture (*Sarcoramphus papa*) has been recognized as a very special being. The Mayas considered it to be a god and believed that it carried messages between other gods and humans. This comes from the codices which are accordion-style folded pages written by scribes on bark paper, and in which were recorded Maya knowledge and religion. Spanish conquistadores and the priests who accompanied them destroyed thousands of the codices; only four survived. The codex that describes the king vulture was one of the four. The magnificent bird was represented with the glyph shown here.

Since it was the king, Orlando and I had referred to the vulture as a "he," but it could just as easily have been a "she." There is no visual difference between the sexes, although males do tend to be slightly larger. I have seen around 15 of these beautiful birds at Hacienda Baru, but, strangely, I have never seen two together. I say strangely because king vultures mate for life. They normally nest in a hollow tree. The female lays a single egg which is incubated by both parents. The chick is nearly helpless when hatched, but does have a little fuzz covering its body. The parents try to keep the nest as stinky as possible by vomiting in and around it. I can only imagine how putrid vulture puke must be. The purpose of this practice is to repel predators, but even so, the parents take turns guarding the smelly nest. I guess it is a good idea for the chick to get used to foul smells considering what it will be eating for the rest of its life. After the chick becomes ambulatory it practices projectile vomiting as a defense against intruders.

I have never been close enough to a king vulture to smell it, so I can't say for sure whether it stinks or not, but I imagine it smells pretty bad. In addition to all the vomiting described above, it has a strange habit of pooping on its own legs. The reason for this is allegedly to reduce its body temperature. Evaporation of the feces covering its legs and feet has a cooling effect. There is even a name for this behavior: urohidrosis. It sounds a little far out to me, but biologists seem to agree about it.

Hacienda Barù guide Juan Ramón told me that when a tough-hided animal dies, like a sloth or coati, the black and turkey vultures have to wait for it to decompose a bit before they can penetrate the thick skin, open the carcass, and get to the goodies inside, but the king vulture has a sharp-hooked beak and is big enough and strong enough to rip it open when fresh. An adult male may weigh as much as four kilos and have a wingspan of two meters.

When I started learning about birds in the early 1990s everybody seemed to be in agreement that king vultures have a very keen sense of smell with which they detect the dead animals that they eat. Their preferred habitat is primary forest where several layers of canopy obscure the ground from a soaring vulture. I once flew over the Osa Peninsula, almost all heavily forested, in a small plane. We were over the forest for about 45 minutes. I was amazed at the number of large white vultures with wings trimmed in black below the plane riding the thermals above the canopy. We must have seen 30 or 40 of them.

In my experience at Hacienda Barú, and in that of our guides, when an animal dies in the rainforest the first vulture on the scene is always the king vulture. Most of my personal sightings have been in the dense forest with only two at the edge of the forest and one in a lightly wooded area. I have never seen one in an open pasture.

Recently some controversy has arisen over the olfactory capability of the king vulture. In fact, the Cornell Lab of Ornithology goes so far as to say that it appears king vultures do not have olfactory sensors to locate carrion by smell. On the other hand, the Smithsonian Zoo says the bird has both keen eyesight and a keen sense of smell that it can use to find its food. Both institutions are highly respected scientific organizations, so we can conclude that more research needs to be done to clear up the doubt. Based on my own personal experience I will, for now, side with Smithsonian.

Though king vultures are seldom seen, they aren't considered to be endangered. They do appear on the IUCN Red List, but are listed as "Of Least Concern." Natural enemies are few, if any. Possibly a large boa would eat a fledgling. Considering the bird's diet, its meat must taste horrible. I have never seen any kind of bird or animal scavange the carcass of a dead vulture. I have never heard of a person killing a king vulture, though I imagine it does happen on occasion. In some rural areas of Costa Rica, the hunters believe that if you shoot a king vulture, your rifle will never shoot straight again and will never kill another animal. All you can do is throw it away and buy a new one. This bit of folklore may serve as a deterrent to hunters.

In 2003 my daughter Natalie and I visited Corcovado National Park. We stayed a couple of nights at La Sirena Station and, on the third day, hiked 22 kilometers to Corcovado Lodge, near Carate. Much of the walking was on the beach. At one point our biologist guide Charlie stopped for a brief rest. "This spot brings back memories," he mused. "About a year ago I came past here, and a dead whale had washed up on the beach right over there by those rocks. I still remember the stink. It was suffocating. The amazing thing was the king vultures. There must've been 40 or 50 of them on that carcass, and not a black one to be seen. I still have that image clearly imprinted in my mind," he sighed. "And you know what? I didn't have a camera."

15. A Passel of Jacks, Jills and Joeys

Pests? Maybe. Interesting? Definitely!

I have seen lots of Jacks and Jills and even a few Jills with bunches of Joeys, but I have never seen a passel. In fact, I just recently learned what one was. I know you have heard of a pride of lions, a herd of elk, and a troop of monkeys, but how about a passel of opossums? Since these mammals are solitary by nature and don't hang out in groups, I doubt if the word sees much use. Jacks, of course, are the males, Jills the females, and Joeys the babies. I don't know if a Jill with a bunch of Joeys on her back is considered to be a passel or not. If it is, then I guess I have seen a passel.

The common opossum (*Didelphis marsupialis*) is a nocturnal marsupial mammal found throughout this part of Costa Rica. They are not the least bit charismatic, and certainly won't win any beauty contests. Being stinky when up close doesn't win them any fans either. The Bribri Indians even consider them to be messengers of death. Not being picky about their food often gets them into trouble with urban dwelling humans for invading garbage cans. Rural families despise them for raiding chicken coops, killing chickens, and eating eggs. They will eat almost anything from fruit and grass to birds, insects, small reptiles and even carrion. With a diet as diverse as that they need lots of teeth of many different kinds. Opossums have more teeth than any other mammal. There are fifty in total, including incisors, premolars, molars and canines.

Despite their bad reputation opossums are really pretty interesting. Let's take, for example, the fact that they are immune to snake bite. Scientists have been searching for the source of this immunity for some time and have recently discovered a chain of amino acids in their blood serum that appears to neutralize snake venom. The really good news is that they have been able to pass this form of immunity on to mice in a laboratory, and there is hope that someday soon it will be available for snake bite treatment in humans.

When I was in school, in about the seventh or eighth grade, our science teacher told us how special humans are because we have an opposing thumb. Because of this superior trait we can do all sorts of things that other animals can't, we were told. As kids we used to joke about how special we were because of our opposing thumbs. What a disappointment when I learned that opossums also have opposing thumbs, but only on their hind feet.

The family *Didelpidae*, which includes eight species of opossums

A Jill with a Joey on her back.

in Costa Rica, are the only family of marsupials in the Americas. Most of the rest, kangaroos, koalas, wombats, and bandicoots among others, live in Australia. This is the only order of mammals to carry their young in a pouch located on the mother's belly. The young, which are more like a fetus than a baby at birth, need to live in the warm nurturing environment of the marsupial mother's pouch for a couple of months before venturing out into the world.

Nobody knows when or by whom it was discovered that a male opossum has a two-pronged penis. This interesting bit of knowledge gave rise to the belief that the Jack mates with the Jill's nose, and when the Joeys are born, she sneezes them into her pouch. I first read about this folkloric belief on the internet and got the impression that it was referring to backwards areas of the United States. Later I found that the belief used to be common among the rural residents in this part of Costa Rica.

The truth of the matter is only slightly less astonishing than the belief. The Jack does have a forked penis, and the Jill has two vaginas and two wombs. A pair of opossums mating appears pretty much the same as any other pair of mammals mating (not in the nose). The Joeys are born only 12 to 14 days after mating. Birth takes place through a third canal that forms temporarily, solely for this purpose, and then disappears.

At birth the Joeys are so tiny, about the size of a pea, that they are sometimes referred to as larvae rather than babies. Nevertheless, they are mobile and able to crawl from the birth canal through the hair on the Jill's lower belly to the pouch where they enter and search for a nipple, of which there are about ten. There is no sharing: only one Joey per nipple. When the larva grabs the nipple, it instantly swells, locking itself into the larva's mouth. There it will remain, unable to let go, for a couple of months until it is mature enough to venture outside. After that the Joeys often ride on mom's back. Around twenty Joeys will be born, but only those who are fast enough and strong enough to find the pouch and a nipple will survive. The others perish.

The oldest marsupial fossils yet to be discovered are about 75 million years old. These mammals are survivors. They were here before the demise of the dinosaurs, and they survived the catastrophe that ended the reign of those huge reptiles. It has crossed my mind that the opossums could be considered the cockroaches of the mammal world. Like those unsavory little beetle-like insects, they are generally despised, will eat almost anything, have been around a long time, have survived great upheavals, and they are still going strong.

And opossums really do "play possum." When threatened, and with no avenue of escape, they will lie still, mouth open, tongue hanging out, slow their breathing, and remain this way for as long as several hours. This may buy them some time and an opportunity of escape.

A Jill overloaded with Joeys.

16. Chocolate Addiction: Not Just for Humans

Lots of Wild Animals Love Cacao Seeds

More than a dozen years ago when I wrote the original article, "Monkeys Are Made of Chocolate" for the magazine *Quepolandia*, little did I know that it would later become chapter one of a book by the same name. In it I talked about the unexpected effect that the increased availability of cacao had on the general health and population of white-faced monkeys.

I wrote that when we abandoned our nine hectares of cacao plantations at Hacienda Barú, due to a market plunge, the monkeys immediately moved in and started chowing down on cacao syrup. As a result of this windfall of nutrition the white-faced capuchins no longer had to struggle to survive. Rather they got fat and happy and started having more babies than ever. At that time, I was only beginning to understand the complexity of a cacao plantation given over to natural processes.

Man can plant cacao to produce chocolate for the exclusive consumption of humans, which all of my fellow chocolate lovers will appreciate, but only Mother Nature can exploit it to its fullest. One of the most researched protected areas in Costa Rica is a huge nature reserve called La Selva Biological Station. A biologist friend once told me that the most biologically diverse areas of La Selva were the old, abandoned cacao plantations, not the primary forest that one might expect. This got me to thinking and paying more attention to our old cacao plantations at Hacienda Barú which have not been pruned or otherwise cared for since 1986. I soon figured out that the cacao-producing trees were planted only four meters apart, and that nowhere in a natural forest do we find such a concentration of nutrients.

Early on during my decades of experience with the Costa Rican rainforest I noticed how wasteful white-faced capuchin monkeys can be. Their normal mode of eating anything is to grab a handful, stuff part of it in their mouths, and throw the rest on the ground. With a cacao pod they pluck the pod, weighing up to one kilo, take a bite out of one side or bang it against a branch to break it open, slurp the syrup from a few seeds, throw the rest on the ground, and pick another pod.

Wastefulness? One of our guides observed a group of monkeys on the ground messing around with one- to two-week-old fallen cacao pods. Upon closer examination he realized that the pods were full of protein-rich maggots which the monkeys were eating. Not as wasteful as I had thought. At first there was little competition for the fallen pods. Later the pacas, agoutis, coatis, spiny rats and peccary discovered them. What is left after all these animals are finished is consumed by ants, fungi, bacteria, and other minuscule forms of life. Absolutely nothing goes to waste. Even the trace minerals found in the cacao pod are incorporated into the soil and are utilized by other organisms, mainly plants.

Monkeys also prey on squirrels and their young, so the squirrels only make occasional incursions into the old plantations, and never nest there. Their reason for being there is to eat the nutrient-rich cacao seeds which they do by chewing a hole in the pod without removing it from the tree, digging the seeds out with their front paws, and eating them. As you might imagine, many seeds fall on the ground while the squirrel is eating. Those seeds remaining in the pod once the meal is finished gradually loosen up and fall out of the hole onto the ground. Again, nothing in the natural world goes to waste. Those animals already mentioned, and several others gobble them up as soon as they find them. Peccaries and coatis rout around in the leaf litter and loose soil, which is rich in organic matter, and eat the seeds, along with many forms of small animal, fungal, and plant life found there.

A pair of slaty-tailed trogons, breathtakingly beautiful rainforest birds, were often seen near an abandoned cacao plantation. Because of their secretiveness it took us a couple of years to figure out where they were nesting.

Termites eat dead wood, but don't bother the living wood. Their nests are large, dark, brown or black globs which are found on many cacao trees in abandoned plantations. When the plantations are being cared for, there is no dead wood. It all gets pruned away, and there is nothing left for the termites. We discovered that the trogons were burrowing into the center of these hard, brittle nests and hatching and caring for their young in a hollowed-out cavity within the termite's abode. Parakeets and some of the smaller parrots also take advantage of the termite nests for the protection of their eggs and chicks.

When a Hacienda Barú guide is conducting a tour that passes through an old cacao plantation, he/she will often eat a termite and offer one to each of the guests. The closest I can come to describing the taste is that it strongly resembles peanut butter. I guess geckos, and a multitude of other little lizards like peanut butter because they frequent the branches where the termites are found, break open the tunnel-like trails, and eat the termites as they scurry around trying to repair the damage to their trails. We have noticed several different species of hawks in the plantations, and upon careful observation, realized that they are there mainly to feed on these small lizards.

Boas are often seen coiled up in the branches of the cacao trees. Their main prey are squirrels, opossums and rats. Sloths don't eat cacao leaves, but they do eat the leaves of some of the shade trees found in the plantations, and are frequently found there. Many of those shade trees are a species called poró which produces hundreds of a nectar-laden, orange flowers. Monkeys love to eat the flowers and hummingbirds are attracted to them like flies to sugar. One ornithologist counted eleven different species of hummingbirds in one heavily flowered poró tree.

Costa Rican wild turkeys called great curassows scratch around in the leaf litter and loose soil looking for seeds, insects, and other minuscule forms of life. Litter frogs and occasionally poison dart frogs can be found on the moist ground as well.

In addition to all of the animals mentioned previously, our trail cameras have captured videos of grouse-like great tinamous, armadillos and also ocelots in the old cacao plantations.

Recently I went with Hacienda Barú guide Cristian into an old cacao plantation hunting for photos for this essay. It was an incredible hike. First a collared peccary appeared like magic in the middle of the trail, looked around and casually wandered back into the trees on the other side. About 15 minutes into the hike we encountered a troop of monkeys feeding in a group of heavily fruited trees. We observed a young monkey pluck a cacao pod and repeatedly bang it against a branch to break it open. Half the pod fell to the ground while the monkey fished for seeds with its fingers in the other half. Shortly another young monkey showed up and tried to steal the half pod from the first. Not surprisingly it fell to the ground during the fracas. With nothing to fight over they each went on their own merry way.

"Look," said Cristian, "a mother and baby agouti." They were happily eating the fresh seeds that the monkeys had dropped. In the lower branches a gray-headed tanager flitted around grabbing the insects that the unruly monkeys stirred from their hiding places. The whole experience clearly illustrates the reasons for my ongoing love affair with tropical nature.

Photo: Average-size cacao pod

17. Strange, Creeper Cats

In the Boondocks, They Call It a "Black Panther"

When I first laid eyes on the two black "kittens," a quote from a Robert Heinlein novel popped into my mind. It has been so many years ago since I read it that I can't even remember which of his novels it was, but I remember the quote. In referring to a complex subject, Heinlein said that making sense of it was, "...like searching in a dark cellar at midnight on a moonless night for a black cat that isn't there."

These two kittens were that black without a hint of any other color. Even their eyes were black. In addition to their extreme blackness there was always an air of mystery about them. They didn't walk like ordinary cats, rather they walked all crouched down, more of a creep than a walk, like they were constantly stalking something. They never made any noises other than purring; they never clawed the furniture; they were never underfoot and never got into trouble of any kind. There was always something strange about them. We named them Hocus and Pocus.

We found them on our doorstep one Sunday morning. Their

eyes and ears were open and they could get around just fine, but they were really too young to be weaned. Diane took them in and bottle fed them until they were old enough to eat solid food. Hocus and Pocus grew quickly and never got sick. They were friendly enough, but never cuddly like most kittens. They always had an air of independence about them that let you know that they didn't belong to anyone.

All cats are that way to a certain extent, but Hocus and Pocus were even more so. They wouldn't reject affection, but they never looked for it. Our grandson Shawn used to drape one of them around his neck and walk around the house that way. The cat was totally non-committal about these antics, neither rejecting Shawn's affectionate game nor craving the attention. Diane said she could tell them apart, but I didn't even try. Like all cats they never gave any indication that they knew their names. I can definitely say that they were the coolest cats I have ever known.

Both Hocus and Pocus were males, and when they were old enough Diane got them their shots and had them neutered. Though only a couple of months old, they were already about the same size of an ordinary adult house cat. A couple of weeks after they were neutered a biologist friend of ours came to visit. He walked into the house, took one look at Hocus and Pocus, and said, "Wow, where did you get the jaguarundis?"

Costa Rica's Only Wild Black Cat

Jaguarundi is the common name of Costa Rica's only wild, black cat. Some jaguarundis are tan to reddish-tan or sometimes gray, but we don't have those color phases in the southern part of Costa Rica. The scientific name is *Herpailurus yagouaroundi*, which has an interesting derivation. The first part of the word is either from the Greek word *herpa*, which means strange, or the Greek word *herpes* which means creeper. The ending of the name is derived from the Greek word *ailouros* meaning cat: strange, creeper cats. The species name *yagouaroundi* is also the common name given to these cats by the Guarani Indians from South America.

Compared with other cats, the jaguarundi's head is small in relation to its body. You could say that of the wild Costa Rican cats, it is the least cat-like. Some people even notice a similarity with certain mustelids like otters or weasels. In fact, I have heard jaguarundis called both "weasel cats" and "otter cats," and they are often mistaken for another mustelid, the tayra. The tayra is a lot like a weasel, but weighs a little over 5 kilos. A jaguarundi is a little bigger, but not noticeably so, and both are black. As most sightings of these two animals are fleeting glimpses of a black, cat-like mammal running across the road or the path, the two are often mistaken for one another. When I spot one, I always look at the tail. The jaguarundi has a typical cat's tail, long and thin, whereas the tayra's tail is not so long and a lot thicker and bushier.

I remember an incident when I was galloping along on a horse, and either an adult tayra or jaguarundi ran across the path about ten meters in front of me. A couple of seconds later two smaller, black mammals also attempted to cross the trail. They ran right under the front feet of my horse. The horse tripped and almost fell but was able to regain its feet. The two young, black mammals went rolling head over heels, landed upright and took off into the bush. The incident happened so fast that I didn't have a chance to look at their tails.

My best sighting of a jaguarundi was in the mid-afternoon one day during the dry season, on one of the trails at Hacienda Barú. I had stopped in the shade of a palm tree to scan the stream for signs of life. The breeze was in my face. A lone jaguarundi stepped into the field of view of my binoculars. I watched it wander around for several minutes, mostly in the partially dry stream bed. It turned over rocks and peeked into hollows in the stream bank. At one point it briefly stalked a small lizard, but the lizard easily escaped. The jaguarundi captured a crayfish in a small pool and ate the whole thing. Finally, the black cat walked out of the stream and into the forest.

Though there was nothing particularly scary or exciting about the incident, it still made my heart beat a little faster. Neither before nor after have I had the good fortune to observe a jaguarundi doing what they do in the wild.

In addition to crayfish and little lizards, jaguarundis eat some fruit and vegetable matter, large insects, fish, rats and other small rodents. They have been known to eat opossums, but don't normally tackle prey that large. Ground nesting birds like tinamous and wood rails form a large part of their diet. Though they are agile climbers, they hunt mostly on the ground.

A Final Word on Hocus and Pocus

Once we found out that Hocus and Pocus were jaguarundis we looked at them differently and were able to see their strange behavior in a new light. Also, we regretted having neutered them. "If only I had known...," lamented Diane.

But the damage was already done, and there was no way of undoing it, so Hocus and Pocus would just have to live with their infertile condition. We assumed that they would continue to live with us like house cats. "They're addicted to the easy life," I remarked. "I doubt they will give up being fed and pampered to go out and fend for themselves." Being an animal under Diane's care could be compared to nirvana, the highest level of reincarnation.

Though Hocus and Pocus had been quite small when they joined our household, the wildness was still deeply embedded in their psyches, and all the luxury living in the world wouldn't change that. About six months after we found out that they were jaguarundis, one of them left. He just walked out the door one day and never came back. Diane says it was Pocus. A month later, almost to the day, Hocus left. We didn't notice he was missing until he didn't show up for the afternoon feeding. Diane kept hoping that they would return, but I knew they were gone for good. The wild instinct was in their blood, and they were where they were born to be. I was happy for them.

Less than a month after their departure, one of our former black pets returned to the house for a final look. Diane spotted him near the gate early one morning. He looked toward the house, looked at Diane, turned and wandered off. About six months later I was walking down a rainforest trail near our house when a black cat stepped into the trail less than two meters in front of me. I stopped dead in my tracks and watched. The cat looked at me straight in the eye. "Hocus? Pocus?" I called. Whoever it was gave no sign of alarm, but neither did it give any sign of recognition. The strange, black cat creeped across the trail and into the forest on the other side. He never looked back.

Photo: Rigo Pereira Rocha

18. Bugs, Bugs, Bugs

Are Insect Populations Really Decreasing?

We didn't have cockroaches in Colorado where I grew up. At least I never saw any. The first time I heard of them was on television. For those of you who are old enough to remember, it was on the show "The Honeymooners" in which a couple named Ralph and Alice lived in a rundown apartment somewhere in New York City. One day Alice was complaining that she had seen a cockroach in the apartment. I asked my mom what a cockroach is and was told that they are horrible bugs that are only found in slums. Later in life I found out that in the tropics they live everywhere. In the chapter titled "La Casona" of my book, *Where Tapirs and Jaguars Once Roamed*, I wrote about how entertaining it was to hunt them in the evenings back in the days before we had electricity and illuminated the house with kerosene lanterns.

Other than butterflies, most people don't like any kind of insect or other small arthropod very much. By other small arthropods I am talking about spiders, centipedes, and other little creatures similar to insects. Bees, wasps, scorpions, ants, and spiders can inflict painful bites. Many kinds of them eat our food, crops, and ornamental plants. Termites even eat our houses. Some insects carry disease. Mosquito-borne malaria kills nearly a half million people annually. Dengue fever, chikunguña, and zika are all carried by mosquitoes. Chagas disease is carried by the kissing bug. Ticks carry and transmit lime disease and Rocky Mountain spotted tick fever, among other diseases. The list goes on.

Ask the average person to name something beneficial that insects do for us, and they will probably have a hard time thinking of something. In reality these small creatures are essential to the well-being of all other forms of life on our planet. Pollination and decomposition of dead matter are two of the most important services they provide, but there are others. Edward O. Wilson, in his classic book, *The Diversity of Life*, puts it this way: "So important are insects and other land-dwelling arthropods that if all were to disappear, humanity probably could not last more than a few months."

If that sounds a bit extreme, consider this: Without insect pollination many species of plants would cease to reproduce and eventually perish. This would be well over 100,000 species (hundreds of billions of individual plants), many of which provide food for humans. All the species that feed on those plants would cease to exist as well. Many species of animals that feed on insects would perish. These include amphibians, reptiles, birds, and mammals.

One example would be insect-eating bats of which it is estimated that in the state of Texas alone, during the summer months, upwards of 100 million bats consume hundreds of tons of insects daily. And that is only in Texas. All those bats would starve to death along with everything else that eats insects over the entire planet. Without insects that feed on dead matter, the remains of these dead plants and animals would not decompose normally. Instead of being broken down into their basic elements in a few days or weeks, the dead material would take months or years to decompose. The planet would become a rotting, stinking mess.

Recently there has been a certain degree of alarm among biologists about declining insect populations worldwide. Some have even called it Insect Armageddon. The best article I have read on the subject is titled "As Insect Populations Decline, Scientists are Trying to Understand Why" by Mary Hoff. It was originally published in *Ensia Magazine* and later reprinted in *Scientific American.* Pesticides appear to be the number one culprit and habitat loss a close second. Most of the evidence is anecdotal, and little is based on research. Scientists seem to agree that insect populations are declining, but a great deal more study is necessary to determine how serious the situation is and the exact causes. I think we can expect to hear more about this in the future.

I first became aware of the problem several years ago when a professor from the Firestone Center for Restoration Biology contacted me and asked if we had noticed a decline in golden orb weaver spiders. She said that some of her students had been studying them for several years and each year they are more difficult to find. Thinking about her question I realized that the golden-tinted webs used to obstruct the walking trails on Hacienda Barú, but that I hadn't seen one for quite a while. My wife Diane says that she used to get someone with a ladder and a long broom to remove all the spider webs from the ceiling about once a month, but not anymore. It seems as if populations of other species of spiders are also declining. Mosquito populations have declined tremendously. Butterflies as well. At Hacienda Barú we are definitely seeing diminishing numbers of some species of insects and small arthropods.

Since I started paying attention to declining insect populations, I noticed something else interesting that is happening around Hacienda Barú. Some species of insects and small arthropods appear to be increasing in number. In all my years of living in Costa Rica, I've never seen so many millipedes and centipedes. They are all over our house, in the lawn, everywhere you look. I thought that maybe this was a local phenomenon only at Hacienda Barù, but I saw something about the millipede plague on social media the other day. It seems to be widespread. Fortunately, they don't bite or sting, and aren't that difficult to put up with.

Another example of insects that appear to be increasing, at least around our house, are ants. Neither Diane nor I remember ever having so many of them in the kitchen. An increase in ants could explain the decrease in butterflies. Ants seek out and eat the butterfly eggs.

Everything is intertwined. We have always had gnats, but I don't remember them being so numerous. A small cloud of these tiny mosquito-like bugs hangs out around my computer; so many, in fact, that I have to be careful not to inhale them.

I think it is much too early to call insect declines an Armageddon, but it is a phenomenon that we are going to hear much more about in the future. Hopefully it will occur to researchers to investigate increasing populations of certain species of insects as well.

And there is still a healthy population of cockroaches if anyone wants to try hunting them.

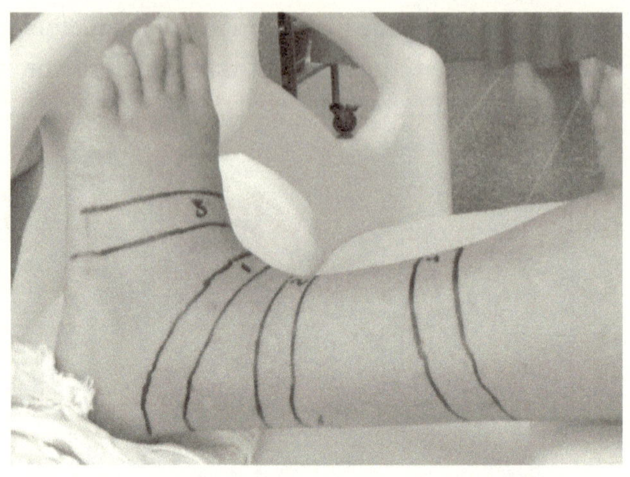

19. More About Snake Bite

A Viper on the Porch

In *Quepolandia* a few years ago, I published an article entitled, "If You Are Unfortunate Enough to Get Bitten by a Snake, Do It in Costa Rica." That issue arrived at Hacienda Barú the same day that my friend and neighbor Randy Burns was bitten by a terciopelo (*Bothrops asper*), Sunday, August 2, 2015. He was awakened at 4:00 a.m. when his dogs started barking. Concerned about what may have triggered the barking, Randy got up and walked barefoot onto the front porch. No sooner did he set foot on the porch when he felt a sting on his left ankle and thought he had been bitten by a scorpion.

Returning to his bedroom, he illuminated his ankle with a flashlight and saw blood. Thinking it strange that a scorpion sting would draw blood he returned to the porch with the flashlight and a stick and shined the beam around the floor. There was the terciopelo coiled up in a corner. He tried to hit it with the stick, but the snake managed to escape out into the yard. "Marie, Marie," he called to his wife. "Hurry! We need to go to the hospital. I've been bitten by a snake."

They called their trusted employee Eliecer, who came running over from his house. Eliecer drove well over the speed limit. They covered the 30 kilometers to the Hospital Max Terán Valls in Quepos in less than 20 minutes. They pulled directly into the area where the ambulances unload patients. Marie jumped out, ran up to the emergency door and told the first person she saw that her husband had been bitten by a terciopelo. They rushed Randy straight into the emergency treatment section ahead of everyone in the waiting area. Marie was impressed that there was no wasted time.

Everyone knew what needed to be done, and they did it. Hospital personnel drew some lines around his ankle and lower leg with a marker pen so they could determine the level of swelling by measuring the distance between the lines. One of the fangs had penetrated all the way to his ankle bone, and the other entered the soft tissue below it. Forty-five minutes from the time he was bitten by the snake, the antivenin started flowing into Randy's vein.

The pain started about the time they got to the hospital. At first it was little more than an annoyance, but once the antivenin treatment started, it almost immediately became unbearable. "Never in my life have I experienced such excruciating pain," Randy told me later. They gave him the most potent pain killers they had, short of morphine, but nothing helped. The pain continued. After six hours of misery the doctor gave him morphine. They first had to wait for the other pain killers they had given him to wear off. The morphine finally brought the pain down to a bearable level.

About half an hour after the antidote began flowing into his veins, Randy had an allergic reaction. A severe rash broke out all over his body, accompanied by an itching and burning sensation. The doctor suspended the antivenin and gave Randy antihistamines to counter the allergic reaction. They waited another 30 minutes and began the antivenin again. This was absolutely necessary in order to save his life. This time the reaction was more severe. Again, they suspended the antidote and gave him antihistamines. His blood pressure plunged and Randy nearly lost consciousness. "I was sure

I was going to die," he confided in me later. The doctor and hospital personnel managed to pull him through the crisis. They suspended the administration of the antivenin permanently. Marie remembers that they had originally injected five vials of the antidote into a half liter bag of saline solution which dripped into Randy's vein. By the time the doctor stopped this treatment for good, about two-thirds of the solution was gone. She figures that he received about three vials. Fortunately, that was enough, even though all five would have been better. Perhaps the snake didn't inject a full load of venom. No one will ever know.

Randy was in the emergency room for two days and was then was moved over to a regular hospital room, where he remained for three more days. He was discharged from the hospital on Friday, August 7. The next night he returned because of high blood pressure. "I thought my head was going to explode," he told me. Twice more he had to return to the emergency room, both times because of excessive pain. One time the pain nearly paralyzed his neck and shoulders. The doctor told him that the powerful antibiotics he had been given to prevent infection from the bite were the cause of his muscle and joint pain. Randy also asked the doctor if many other patients experienced an allergic reaction like his. The answer was that only about one case in a hundred is complicated by the allergy.

Once he was home the pain continued to bother him, especially when he walked. When his leg was elevated, the pain diminished. The rash and itching from the allergic reaction also persisted. Seven weeks later, when I interviewed him, Randy was still experiencing muscle and joint pain from time to time. At one point, patches of superficial numbness appeared at different places over his body. He was still experiencing numbness in mid-October more than two months after he was bitten.

The total bill for two days in the emergency room, three days in the hospital, the three follow-up visits, all of the medicines, and the doctors, came to about $7,500. Compare this to Mr. Eric Ferguson who reported to the emergency room of a hospital in

North Carolina. His snakebite wasn't too serious, and he needed only four vials of antidote. He was released after 18 hours. His bill was $89,277. He had been charged $20,000 per vial for the antivenin and $9,277 for the emergency room care.

Fortunately, the Instituto Clodomiro Picado in Costa Rica produces enough antivenin to supply the demand in the country and export serum to the rest of Central America, Panama, Ecuador, and Nigeria. They are even collaborating in the production of an antidote for the bite of the taipan of Papua, New Guinea. The quality of their product is excellent as is evidenced by the death rate of less than one-half of one percent, one to two deaths of the approximately 600 victims of snakebite in Costa Rica annually.

According the the World Health Organization (WHO) 1.8 million snakebites occur annually world wide resulting in 94,000 deaths and about three times that many amputations and patients with permanent disability. Most of the deaths are people from rural areas in third-world countries who don't have access to treatment. Each year in Africa 30,000 people die from snakebites. In comparison only about 11,500 people died during the last Ebola epidemic.

Though the WHO is aware of the problem, they are doing little to help. For example, they issue guidelines for the treatment of 17 different medical disorders including dengue and sleeping sickness, but don't offer guidelines on the treatment of snakebite which causes more deaths than all of the 17 combined. This means that if a person is bitten in rural Africa, they are unlikely to receive treatment. Even if the closest medical facility happens to have the antivenin, the medical personnel will probably not have the knowledge necessary to administer the treatment safely and effectively. According to the non-profit organization Doctors Without Borders, about 95 percent of the snakebites in Africa take place in rural areas and only about 10 percent are treated.

The best antivenin for African snakes is a product called Fav-Afrique which used to be manufactured by a French multinational called Sanofi Pasteur. However, they quit producing it in January

2014 because it wasn't profitable. The company cited competition from other products as the main problem. Fav-Afrique, when it was available, sold for between $250 and $500 per vial, too expensive for rural medical centers which attend most snakebite victims. Competing products are being produced in Spain, India, UK, South Africa, Costa Rica, and Mexico. According to Doctors Without Borders, only the UK product and the Costa Rican product, called Echi Tab Plus, give good results. I don't know how much the Costa Rican product exported to Nigeria costs, but the Polivalent Serum produced by the Instituto Coldomiro Picado costs Costa Rican hospitals about $20 per vial. This is proof that developing countries can produce excellent treatments for the bites of venomous serpents at a reasonable cost.

After Randy Burns was bitten by the terciopelo, he and Marie received several phone calls from friends and neighbors. Many of these calls came from campesinos who warned them that the snake would return in seven days. Eliecer hunted for it diligently all week to no avail. On the morning of Saturday, August 8, Eliecer's mother-in-law called to warn them to be very careful because the snake that bit Randy would return on that very day. Later in the morning Eliecer found it in the car port under a hub cap and killed it.

20. We Certainly Didn't Kill Any Orangutans

We Did Produce Lots & Lots of Vegetable Oil

Most of our clients at Hacienda Barú Lodge come here by way of the coastal highway which takes them through all of the oil palm plantations beginning in Parrita and extending a total of 40 kilometers all the way to Portalón to the south of Quepos. I often hear comments such as: "Aren't all of those oil palms horrible. I hear they are really bad for the environment, and somebody told me that palm oil causes all kinds of health problems." In recent years oil palms and palm oil have gotten a lot of bad press and an extremely bad reputation, some of which is deserved and some of which isn't.

If you search online for "oil palm plantations," many websites will be about the demise of the orangutan due to expanding deforestation in Indonesia and Malaysia. During the decade from 2000 to 2010, an average of nearly one-half million hectares of orangutan rainforest habitat was destroyed **each** year. The total area dedicated to palm oil production in those two countries is now upwards of

8 million hectares. If the deforestation continues it could very well bring about the extinction of these large, charismatic primates in the not-too-distant future.

So why are Indonesia and Malaysia so obsessed with producing palm oil? As you may have guessed, it is a very profitable business. Up until a couple of years ago oil palm was the second largest source of vegetable oil in the world, right behind soybeans. The trees begin to produce the oil-bearing palm fruits within two to three years of planting. Several years later they reach full production and can produce for up to 30 years. A plantation in full production can yield up to 30 metric tons of fruit per hectare per year.

Each hectare of oil palm will produce up to ten times more oil than one hectare of soybeans. You then might wonder, "Why does anyone continue to grow soybeans?" Illinois is the largest soybean oil-producing state in the US. If Illinois farmers could grow oil palms they probably would, but outside of a greenhouse you aren't likely to see any palm trees in Illinois. They don't do well in most of China either, and China is the world's largest soybean producing country.

USDA estimates for 2014 show that Indonesia and Malaysia combined will produce 55,000,000 metric tons of palm oil, more than all the rest of the world combined and more than the entire world's production of soybean oil. We are talking about a $40 billion per year business which is growing rapidly and consuming orangutan inhabited rainforest as it does so. Is it any wonder that palm oil production has a bad name?

Oil palm trees produce large clusters of palm fruit, each weighing as much as 100 pounds and containing 1,000 or more fruits. Each fruit has a kernel or seed in the center. This is covered with a fleshy pulp, and the entire fruit is enclosed in a thin fibrous outer skin. Both the pulp and the kernel are rich in oil. The oil produced from the pulp is rich in carotene, giving it a reddish color. Some sources say that it contains 16 times more carotene than carrots. This oil is high in saturated fats and is therefore solid at room temperature. For this reason, it is ideal for products such as margarine and certain cosmetics.

Oil Palm in Costa Rica

In 2020 Costa Rica produced 270,000 metric tons of palm oil making it the third-highest producer in Latin America behind Columbia and Ecuador. It is possible that some rainforest was cut to make way for oil palm in this country, but, if so, it was done illegally. The land now occupied by the plantations in the Quepos-Parrita area was cleared for bananas many years before anyone in Costa Rica ever heard of oil palm.

Don Oscar Monge Maykall, in his wonderful book *La Historia Real de Quepos*, recounts the history of banana production in Costa Rica. It all started in 1870 when a fishing boat owned by Capitan Lorenzo Dow Baker bought 160 bunches of bananas in Jamaica for one schilling each and took them to Boston where he sold them for $2 each. This strange new fruit rapidly became popular in the United States, and the demand soon exceeded the supply. Dow Baker partnered with Boston businessman Andrew Preston to develop a market for bananas in Boston. The partnership prospered and in 1885 they took on ten new partners and founded the Boston Fruit Company.

The business grew with the demand, and by 1888 a total of 16 million bunches of bananas were imported into the United States, and the market was demanding more. In 1899 the Boston Fruit Company merged with several companies owned by industrialist Minor Keith who had spent the previous decade building a railroad from Port Limon to San Jose in Costa Rica. The United Fruit Company was born of this merger. The company bought land throughout the Caribbean, Colombia and Central America for the production of bananas. Keith received concessions on large tracts of land along the railroad. These he cleared and planted to bananas.

By 1908 about 14,500 hectares of bananas had been planted on the Caribbean side of Costa Rica. A devastating fungus called Panama Disease and later another disease, Sigatoka, destroyed so many banana plantations that by 1931 only 3,500 hectares remained.

In 1923 investor Agathon Lutz Stiegle surmised that the Central Pacific coast might meet the requirements for growing bananas. The annual rainfall is about the same as the Caribbean side of the country, but there is a well-defined dry season during the first three months of each year. Perhaps this type of climate would help control the fungus. He acquired and cleared some land near Parrita and began growing bananas. The experiment was successful and Lutz continued to acquire and clear land on which to cultivate bananas. A small port was built in Quepos to facilitate shipping and the bananas were transported from the plantations to the port on rails with carts pulled by mules and horses. In Spanish a railroad is a *ferrocarril*, but this form of transportation they called the *burrocarril*.

In 1934 the United Fruit Company acquired Agathon Lutz' extensive holdings and continued acquiring land and expanding banana production. A larger, better equipped port was built in Quepos and began operating in 1939. All during the 1940s land was acquired, cleared and planted to bananas. A narrow-gauge railroad was built to transport the bananas from the plantations to the port. It extended all the way to the village of Portalón to the southeast of Quepos. People from all over Costa Rica and Nicaragua migrated to the area to work in the plantations. The area prospered.

In 1954 torrential rains caused rivers to overflow their banks and flood the United Fruit Company's banana plantations. Those that were destroyed by the floods were replanted, but were destroyed again the following year when even more devastating torrential rains flooded a larger area. The railroad bridge over the Savegre River was carried away by the flood waters. The extremely humid conditions created by the floods enhanced the growth of the same fungal diseases that had devastated the banana plantations on the Caribbean side of the country earlier in the century. This brought an end to the production of this popular fruit in the Quepos-Parrita area.

The United Fruit Company had been experimenting with oil palms for a number of years and decided to try planting them in the Quepos area. The palm plantations did well, and the production of palm oil was profitable. In the 1990s the operation was acquired by Palma Tica, a subsidiary of Chiquita Banana. Palma Tica is now part of the Numar Group. They have been successful in producing palm oil throughout Costa Rica. They have developed several hybrid varieties of oil palms that don't grow so tall as the original species. This greatly facilitates harvesting.

Bad for Your Health? Bad for the Environment?

I have heard a few comments that palm oil is bad for your health. Whenever anybody tells me that something is bad for my health I always remember the big cranberry scare in the mid-1970s. The USDA warned the public that many cranberry producers had used a chemical that causes cancer. This information was released in the fall just before the holidays when cranberry sauce has always been a traditional part every Thanksgiving and Christmas dinner. Most of the crop rotted as the public broke with tradition and did without the delicious red fruit for the holidays.

Needless to say, quite a few cranberry producers filed for bankruptcy. Several months after the holidays, all of the science on the matter was completed. The results showed that if a person ate two tons of cranberries a day for a couple of years there was a 50 percent chance that they would develop cancer. Since that time I don't worry much when people tell me something is bad for my health.

Palm oil is high in saturated fats and may cause cholesterol levels to rise more than other vegetable oils do. However the adverse effect is much less than with meat and dairy products. Anyone who has had problems with high cholesterol should think twice before switching from margarine, made with palm oil, to butter.

It has been said that oil palms are bad for the environment. This is certainly true, but any agriculture other than pure organic farming is going to have some negative impact on the environment. Most of the arable land in this area is either planted to oil palms or rice, and I believe that rice production is far worse for the environment than the palms.

Of course many of us would prefer natural rainforest to any kind of agriculture, but we must recognize that there are seven billion people in the world and they all have to eat. As long as there is a need for vegetable oil in people's diets, oil palms are likely to be around. Most of the plantations in Costa Rica were planted on land that was previously dedicated to some kind of farming. Rainforest was not sacrificed to make way for the palms, and we certainly didn't kill any orangutans.

21. Locally Extinct Little Red Deer

People Shot Them All

In Spanish they are called *Cabros de Monte* or simply *Cabritos*, meaning Jungle Goats or Little Goats respectively. Non-forked antlers in the males resemble goat's horns. Red brocket deer (*Mazama americana*) are locally extinct in much of their former range. During my 50-plus years of living in the country, I have seen one and that was in the Corcovado National Park. I believe they are found in Carara National Park as well. The old timers from the south Pacific zone of Costa Rica remember the days when there were many of them.

One story I have heard about hunting brocket deer in the 1950s tells of a hunter who was driving from San Isidro to Dominical and saw a group of the little deer near the village of Barú. Stopping his jeep, the hunter retrieved his rifle from the back seat, leaned it across the hood, and shot every single one of the deer, about twenty as the story goes. Considering that the story is probably an exaggeration, I think there is enough truth in it that we can assume that 60 to 70 years ago there were a lot of brocket deer in the area around Hacienda Barú and that large numbers of them were killed by hunters until the population dwindled down to nothing. I came to Hacienda Barú in 1972 and only heard mention of brocket deer from older friends, employees, and neighbors. When they were mentioned, it was always in the past tense. A visiting biologist saw a buck brocket deer in 1996, and one of the Hacienda Barú guides saw one that same year in the same area. It was possibly the same animal. I saw some scat and footprints about that same time and in the same part of the hacienda. No brocket deer, scat or footprints have been seen since.

I started working with trail cameras about 15 years ago and have never captured a photo of one. The environmental organization ASANA monitors wildlife with trail cameras throughout the area within the Path of the Tapir Biological Corridor which comprises a strip of land about 80 km long and 15 km wide along the coastal ridge between the Savegre River and the Térraba river. During their five years of monitoring with cameras, they have never captured a photo of a brocket deer. I think it is a pretty good bet that the species is locally extinct.

There is a small but stable population of white-tailed deer in the biological corridor, so why wouldn't there be any brocket deer? The two species feed on different things, white tails mostly on leaves, and brockets mostly on fallen fruit and seeds, so there isn't much competition for food, and both types of sustenance are plentiful. Though there is a slight difference in size—30 kilos for the white tails and five kilos less for the brockets—that should not necessarily be an advantage. In fact, the smaller, straight-antlered brockets have the advantage of being able to lower their heads and barge through thick vines and undergrowth. Brocket deer do have a fatal habit of running a short distance when startled and then freezing, making themselves a perfect target for a hunter with a gun or an easy catch for a pack of hunting dogs.

The two main causes of local extinction of mammals during the last century have been deforestation and hunting, and both seem to

have played a role in eliminating the red brocket deer from much of its former range.

A biologist friend of mine gave me the photos shown here, which were captured by a trail camera in the Amistad International Park. His 80 cameras located in southern Costa Rica support his ongoing large mammal research project. "Only the cameras at higher altitudes capture photos of brocket deer," he told me. All the sources I have checked give the maximum altitude for the species as 2,800 meters, yet these photos probably came from over 3,000 meters. The executive director of ASANA told me the same thing, "The brocket deer only appear at the upper end of their range and above," she assured me.

In the last 30 years much of the Path of the Tapir has been rewilded and secondary forest now covers much former pasture and farmland. Populations of some species, such as tapirs and jaguars, that a few short years ago seemed to be on the road to extinction, are now markedly increasing. Tapirs have been sighted within the corridor near the northwestern end. Hopefully red brocket deer will join them in their path to recovery.

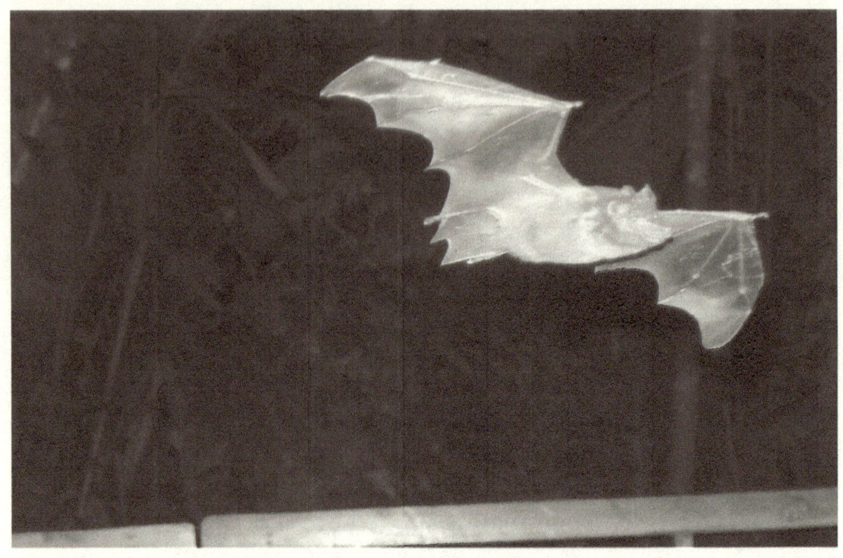

22. Flittermouse?

Amazing Diversity of Our Flying Mammals

What looks like a mouse with wings and flitters around the house? Of course, it's a bat. The name for "bat" in old English was "flittermouse," I imagine because they somewhat resemble a mouse, and their wings flutter or flitter. In middle English it was "bakke" which evolved into "bat" in modern English in the late 1500s.

What comes to mind when someone mentions bats? Probably nothing good. They have been associated with everything from witchcraft, darkness and Halloween to vampires and death. Terms like dingbat and batty have been used to describe people who are foolish or silly, or you might hear it said that a person who acts goofy has *bats in their belfry*.

If there are any terms or sayings that show bats in a positive light, I have never heard them. These sayings and ideas all come from a time when superstition ruled people's lives and science was in its infancy. Today we have a wealth of scientific knowledge about bats, much of which is truly fascinating, yet it is virtually unknown to the public. For example, bats comprise over 20 percent of all mammal species on our planet and are the only flying mammal.

According to Wikipedia there are 5,416 mammalian species on earth and over 1,200 of them are bats. In Costa Rica, and I suspect in most tropical countries, over 50 percent of mammalian species are bats. The smallest mammal in the world is a bat, Kitti's hog-nosed bat, also known as the bumblebee bat, which weighs a mere two grams, about the same as two paper clips. The fastest mammal in the world is also a bat, the Mexican long-tailed bat, which has been clocked at 160 kph, much faster than a cheetah which has a top speed of 120 kph.

I have been around chiropterologists, or bat specialists, for 30 years, and have always been impressed by how passionate they continue to be about the subjects of their study even after years of what appears to be much the same thing. This is because bats are truly charismatic animals and their diversity is extraordinary. In temperate climes like North America and Europe, all the bat species are insectivores. In Costa Rica and other tropical countries, bats eat fruit, nectar, pollen, fish, frogs, blood, and even other bats. I've already mentioned the smallest bat, but the largest New World bat (as distinct from the flying foxes) weighs nearly 100 times more, at about 190 grams. This is the carnivorous species known as the spectral bat which is found at Hacienda Barú, among other locations in Costa Rica, and feeds on birds, small rodents, and other bats.

You may have heard the expression, *blind as a bat.* No species of bat is blind, but vision is not their most useful sense; hearing is. Navigation and food location are accomplished with a process called echolocation which involves emitting a high frequency sound which reflects off nearby objects. The reflection is heard by the bat and allows it to "visualize" its surroundings much in the same manner as radar allows us to detect flying objects we can't see. The emitted signal is so loud it would damage the bat's own eardrums except that special muscles close the ear canal the moment the sound is emitted and open it again in time to hear the echo. Some species, like frog-eating bats, listen for the distinctive sound of their prey in order to locate and kill it. Some biologists believe that this bat can even determine the species of frog from its song and avoid the poisonous or bad-tasting ones.

One thing that most people do know about bats—again it is something negative—is that some of them carry rabies. However, this belief is often exaggerated to sound like all bats have rabies. Actually, very few of them do, the same as very few dogs have rabies. According to the Center for Disease Control, only 6 percent of weak or sick bats that have been captured and are tested for rabies carry the disease. The percentage would certainly be much lower in healthy bats. We often think of rabies when we think of so-called vampire bats. That name is a misnomer as these bats don't suck their victim's blood, and they almost never prey on humans. Sleeping wild and domestic animals are their normal prey. A small cut is made with an extremely sharp tooth, and the bat then laps up the blood that flows from the cut. Their saliva contains an anticoagulant that causes the blood to flow freely.

In all my years of living on Hacienda Barú I've only heard of two cases of blood-eating bats preying on humans. Both were during the years when large numbers of cattle were present in this area, and this abundant source of blood for the bats resulted in unnaturally large populations. A few of them turned to humans for food. Neither victim contracted rabies as a result of being bitten. A mosquito net is plenty of protection against them.

So, while good things about bats are practically unheard of, there are many. For example, scientists are studying desmoterplase, the anticoagulant found in the saliva of blood-eating bats, to determine if it can be synthesized and used in humans for the prevention of heart attacks and strokes.

Bats eat unbelievable amounts of insects of all kinds including mosquitoes that plague humans and many others that attack crops. Their role in pest control means lowered need for toxic pesticides in agriculture.

Seed dispersal is another important service they provide. Once we wanted to plant a nursery of a tree called the ojoche, but couldn't find any seeds under the trees. Later we discovered that bats were collecting the seeds and carrying them to an old cacao plantation where they perched on the low hanging branches of the cacao trees, ate the fleshy pulp, and dropped the bare seeds on the ground where they germinated.

Pollination is another service bats provide for certain species of plants where they feed on the nectar from the blossoms. Bat feces, called guano, is very rich in nitrogen and makes a great organic fertilizer. It is mined commercially from caves where bats have roosted for centuries. I used to climb into our bat-inhabited attic with a whisk broom a couple of times each year to sweep up all the guano to fertilize the gardens around our house.

I almost forgot to mention that bats are some of the most beautiful mammals in the world. And some of the ugliest.

23. The Mixed Feeding Flock

A Magical Rainforest Phenomenon

My obsession with tropical nature began with my first monkey sighting. I had seen them in the zoo, but there is no comparison between zoo monkeys and wild ones. My first toucan was another special moment: "How can it even fly with that huge beak!" My first sloth, boa constrictor, kinkajou, king vulture, and terciopelo viper were others. But it wasn't until 15 years later that I witnessed one of the most complex and intriguing phenomena the rainforest has to offer, the mixed feeding flock: a congregation of many different species of birds in a small area, each busy consuming its own special food in its own special way at its own level of the forest.

Though the rainforest had already captivated my soul a number of years earlier, and I lived in a bird watcher's paradise, it wasn't until 1987 that I got interested enough in our feathered friends to start actively looking for them, identifying them, and making a list. In those days there were no field guides to the birds of Costa Rica.

Birds of Mexico and *Birds of Panama* were as close as we could get. My bird list grew rapidly at first and then gradually slowed. Then on one special day I added eleven new species in less than an hour. About six months into my birding "career" I wandered into a bird watcher's dream, a mixed feeding flock. It is a phenomenon that every bird-watching guide and every birder prays for. For the aficionado it will assure new species for the list, and for the guide it is sure to result in a group of very happy visitors. It goes something like this...

From its perch in the top of a rainforest tree, the toucan grabs a small yellow fruit in its beak, tosses it in the air, and catches it in its open mouth. At that moment a katydid hops by and the toucan grabs and eats it as well. The rainforest is alive with small life forms running in every direction, trying to escape the voracious, foraging army of ants fanned out across the forest floor, stinging, killing and eating every small insect, reptile or amphibian in its path. Some of the ants are climbing the trees *(photo)*, and the toucan will soon have to move to avoid painful stings, but for now it is feasting on every panicked creature that scurries past. A multitude of other birds are also taking advantage of the chaotic scattering of life—hawks, ant shrikes, flycatchers, leaftossers, foliage gleaners, trogons, wood creepers and more, each catching and eating every living thing it finds.

Nothing causes more havoc in the rainforest than an enormous swarm of over 100,000 foraging army ants (*Eciton burchelli*). The presence of raiding ants strikes panic into all types of small animal life, and the birds take advantage of this. Almost never do they eat the ants that create the windfall of food.

The flock contains many different species, each with its own specialty. Any individual or pair of birds from a new species is welcome to join the feast. The newcomers have their own specialty and create no competition for others in the flock. But if another of the same kind as one already present tries to join, they will find themselves unwelcome and repelled by their kin.

This may seem strange to us, but not to the birds. Members of their own species are competitors, but members of other species feed in different manners or at different levels of the rainforest, and there is little or no competition among them. For example, a woodcreeper that climbs up tree trunks, poking its beak in every little nook and cranny, under every piece of loose bark in search of insects, doesn't compete with the leaftossers that pick up leaves on the ground, throw them into the air, and grab whatever moves underneath.

A troop of foraging white-fronted capuchin monkeys which, like the army ants, will cause panic among smaller denizens of the rainforest, will also attract a variety of bird species. As the level of pandemonium is somewhat less than with the ants, the monkey-provoked mixed feeding flock usually consists of fewer birds. The key word is "foraging." Resting, playing, or traveling monkeys don't attract birds. Likewise, traveling army ants don't attract them either. Only groups of raiding predators that are actively searching for food and creating havoc among the thousands of tiny creatures that make their homes in the multitude of hiding places found in the rainforest will attract the birds.

I know of no way you can actively look for a mixed feeding flock. They appear where you find them and never cease to fascinate any human observer who happens to be lucky enough to be in the right place at the right time. I wish you good luck, and sincerely hope you have the opportunity to experience this magical rainforest phenomenon.

24. Killer Monkeys

Eating Was Their Prioirty. Killing Could Come Later.

Monkeys in the zoo are fun to watch but observing them in the wild is almost like watching an entirely different species in a similar body. My first wild monkey sighting was a very emotional experience, and each subsequent encounter with white-throated capuchin monkeys (*Cebus capucinus*) nudged my curiosity a bit further. Even today, decades after that first glimpse of a black-and-white primate, making eye contact with it and watching it scamper up a tree, they still fascinate me. Capuchins are always coming up with surprises and exhibiting never-before-seen behavior. Later I was to learn that they are the most intelligent of Costa Rica's four monkey species.

In the 1970s and '80s written information about Costa Rica wildlife was almost non-existent. There were no bird or mammal books to be found, no internet, and not even any laminated information sheets with photos. For the next 18 years, until the first field guide, *Neotropical Rainforest Mammals*, was published, personal observation, as well as listening to campesino friends, neighbors, and workers provided the knowledge I eagerly sought.

I observed the capuchins playing, fighting, foraging for fruits, flowers, and buds. I watched them in the trees, on the ground, and in the shallow water. Everybody told me that they could be vicious, but I really didn't believe them. Threat displays were commonplace, but nothing I would call vicious. The only somewhat aggressive behavior I had observed was a troop of monkeys harassing a three-toed sloth. It was obvious they were being very careful to avoid the sloth's powerful, grasping claws, but I took this incident to be more of an amusement for the monkeys than a harmful act. I figured that maybe a female could be vicious if another animal threatened her young, but nothing more. I decided that the word "vicious" was probably an exaggeration.

Then one day the workers returned from fixing a fence near the mangrove and excitedly told me of a thrilling experience they had just had. Three capuchin monkeys captured a mature, egg bearing, female green iguana, ripped her open and devoured the eggs in her belly. Then they started on her flesh, occasionally sharing scraps with the rest of the troop. At some point during this ordeal the iguana died. Once the monkeys had her in their grasp, they weren't concerned with taking time to kill her. She wasn't going anywhere, and eating was their top priority.

Upon hearing about this experience, I realized that the campesinos hadn't been exaggerating at all. Monkeys really could be vicious. Though I had always been fascinated with these wild primates, I had never really observed the details of their behavior. From that day forward, every time I came across a troop I paid close attention to their actions. It soon became obvious that vegetable matter was only part of their diets.

A couple of days after the iguana incident I observed a monkey catch a grasshopper and pop it into his mouth. Several months later I saw two of them destroy a bird's nest and eat something from inside, though I was unable to see if it was eggs or hatchlings. At some point I saw a young capuchin chase something up a branch, finally snatching it off the limb with his hand. When he paused briefly to admire the prey before stuffing it into his mouth, I could see that it was a small lizard. One of the workers told me he had seen two monkeys on the ground digging maggots out of a rotten cacao pod and eating them.

It was becoming obvious that white-faced capuchin monkeys were far from the peace-loving vegans I had imagined them to be. They would eat any source of animal protein that presented itself, and if that required being vicious killers, so be it. Thinking back on it now, I realize that their prominent canine teeth, often visible during threat displays, should have given me a clue.

With the onset of the environmental movement and ecological tourism in Costa Rica, I often met field biologists who shared stories about their work. One of the most interesting came from a couple, both biologists, from Guanacaste. Mammals of the tropical dry forest was their field of study, and white-nosed coatis their specialty. Coatis build nests in trees by lining crevices with leaves or by constructing nests of leaves in thick clumps of branches. The young are helpless for the first month of life and are unable to leave the nest until about six weeks of age. Predatory birds and snakes will occasionally take an infant coati, but white-throated capuchin monkeys are the primary predators. In fact, in the tropical dry forests of Guanacaste, the predation became so devastating that biologists feared the coatis would become locally extinct.

What happened next is an excellent example of natural selection at work. After losing their young to the capuchins, several coati females successfully mated again and had a second litter, this one during the rainy season when the humidity and precipitation create conditions that are less than ideal for rearing young. Despite this problem there was a tremendous advantage to be gained. During the rainy season the monkeys have plenty to eat and don't invest as much time and energy searching for coati nests. As a result, a much higher percentage of the litters survive. Within a few years, new generations of females, who themselves had been born during the rainy season, began reproducing, and like their mothers, they gave birth during the rainy season. This change of behavior enabled the species to survive local extinction.

Monkeys and coatis have a strange relationship. Although the former are the primary predators of the latter, I have seen them together in a royal palm tree, side by side, happily feeding on palm flowers. One researcher at Hacienda Barú observed a capuchin monkey grooming a coati that was lying on a branch in a tree where a group of each species was resting peacefully. Yet on another occasion, Hacienda Barú guide Pedro Porras and a group of visitors observed four capuchin monkeys mob a lone coati and throw him out of a tree where he fell through some lower branches and all the way to the ground, 10 meters below.

Knowing of the predation by capuchins of the white-nosed coati nests, I began to wonder about squirrels. Like coatis, their nests are found in the treetops, but are much more visible. Surely the capuchins prey on their young as well, I reasoned. Upon pondering this situation, I realized that the only squirrels I had ever seen in forested areas frequented by monkeys were the pygmy squirrels, not the more common variegated squirrels which seemed to shy away from monkey habitat.

Coatis at Hacienda Baru

Nobody at Hacienda Barú has observed capuchin monkeys and variegated squirrels in the same habitat at the same time,

and there is almost no overlap of their territories. Fortunately, we now have an excellent field guide available, *The Mammals of Costa Rica* by Mark Wainwright. Wainwright tells us that capuchin monkeys have been known to "…team up to chase squirrels, often knocking the squirrels to the ground, and then pouncing on them..." Though Wainwright doesn't say so, we can presume that they then kill and eat the squirrel.

As we have seen, the campesinos were right; white-throated capuchin monkeys can be very vicious. This is not to condemn them; all wild animals must eat to survive. When their food is animal protein, some other animal must die. Fortunately, I have never heard of monkeys attacking humans, though I'm sure that there have been incidents, especially when people feed monkeys. They tend to attack things much smaller than humans, things they can easily kill and eat. Had I had all of today's wonderful information sources all those years ago, I wouldn't have gone so long with my misconceptions about these amazing primates. But then I wouldn't have had so much fun learning.

A big thank you to biologist guide Rigoberto Pereira who captured an amazing photo of two monkeys eating an immature green iguana. That photo was the inspiration for this article.

25. Life Before and After Janet

Bye, Bye Bananas

When the paths of powerful storms take them near the Caribbean coast of Costa Rica, the accompanying low pressure systems draw in weather from every direction, including across the country and out into the Pacific Ocean. As clouds move toward the Atlantic storm, they are halted by a formidible obstacle, the Talamanca mountain range. The moisture-laden air rises in an effort to clear the mountains, then cools, condenses and falls in torrents on the Pacific side of the country.

One recent example: tropical storm Nate caused millions of dollars of damage, 11 deaths, and 11,700 people displaced from their homes. Since I have lived on Hacienda Barú, we have on occasion experienced torrential rains caused by Atlantic hurricanes: Joan in 1988, Cesar in 1996, Mitch in 1998, Stan in 2005, Nate in 2017, and many lesser storms.

From what I've been able to learn from people who were living on the Pacific coast in 1955, all of these storms were mere thunder showers compared to Hurricane Janet. She slowly made her way across the Caribbean on a course similar to that of Nate with the difference that Janet was a Category 4 hurricane, not a tropical storm. Those of you who experienced Nate can only imagine the fury of

Janet. On her trek northwards along the coast of central America and into southern Mexico, she became the first Category 5 hurricane in the history of the region and the first to cause over 1,000 deaths (according to Wikipedia).

There aren't many people alive today who remember Janet, and none of those I have interviewed remember her name. The flood is referred to locally as La Llena de 55, or the 1955 flood. Those who lived through it didn't have access to television, satelite images, early warnings, or internet like we do today. At that time the only way that meteorologists could determine the category of a hurricane was for brave aviators to fly a plane into the eye of the storm and measure the wind speed. One such plane was lost along with 12 crew members gathering data on Janet.

Information about the phenomenon was not readily available to the majority of the people who were ravaged by its fury. All they knew was that torrential rains began to fall and didn't stop for six days. Alejandro Gutiérrez is one of those who lived through the flood, and he never knew that the name of the lady who brought such devastating change to his life was Janet.

Born in 1920 in Río Sequito de Santa Cruz, Guanacaste, one of poorest areas in the country, Alejandro quit school at 9 years of age and went to work to help support his mother and siblings. For his hard work he earned a pittiance of less than $0.05 a day. Life was very hard in Río Sequito, and opportunity was non-existent. By his late twenties he was earning about $0.35 per day swinging a machete for upwards of 10 hours and with no possibility of escaping his impoverished life situation. That's why, when opportunity called in 1949, he didn't think twice about leaving everything he had ever known and going to work for the United Fruit Company in the banana plantations near Quepos. That opportunity presented itself in the form of a recruiter for the company who came to Río Sequito offering work. Alejandro didn't hesitate a second. "Put me on that list," he said. "I'm going to work in Quepos and earn some decent money."

On departure day Alejandro took all of his earthly belongings, a blanket, a few items of clothing, and, of course, his machete. He didn't own a pair of shoes. He and a few other young men walked 40 kilometers to the small port of Barco Quebrado where they boarded a boat called Rio Grande, owned by the banana company. Remaining behind in Río Sequito was his mother, several brothers and sisters, and three illegitimate daughters. Even though Alejandro had only the equivalent of $0.27 in his pocket, he was no stranger to hard work and was full of self-confidence. Quepos and the United Fruit Company were his new life. He never looked back.

Alejandro's first job was clearing overgrown lots for planting experimental crops. There had been problems with fungal diseases in the bananas and the company was looking for possible alternatives, like cacao and oil palm. He was paid by the hectare rather than by the hour. This was a big advantage for a hard worker like Alejandro. He and a friend set out to clear one hectare per day, and they accomplished their goal every single working day. They were paid $6.44 for each hectare, $3.22 for each man, more than nine times what he was making in Río Sequito. For the first time in his life Alejandro had spending money in his pocket.

Later the company moved Alejandro to Damas to work planting cacao. There he met the love of his life, Dominga Gómez. When they began living together in common law marriage, the company provided them with a house. The traditional United Fruit Company village consisted of up to 20 two-story houses built around a soccer field. The houses had a fire box called a fogón for cooking on the first floor and sleeping quarters on the second floor. The couple lived in this type of house for the next six years. Their first daughter was born at a farm called Finca Pastora. Over the following years they lived at several farms: Finca Negro, Finca Mona, Finca Rey, and Finca Marítima, all of which produced bananas. Two more daughters were born while Alejandro and Dominga lived at Finca Rey.

In late September of 1955 Alejandro and his family were living at Finca Marítima located a little more than a kilometer from the

Savegre River. One day rains like no one had ever seen before began to pour out of the sky, and it didn't take long to realize that serious flooding was on the way. The soccer field became a lake, and the water soon rose to the level of the first floor of the dwellings. Alejandro moved the fogón to the second story of the house along with all of their food and other essential items. The waters rose for five days reaching to within half a meter of the second floor before the rains abated and the flood waters began to recede. Everyone was safe and sound. The house was still standing and the family had managed to save most of their possessions.

The company didn't fare so well. The banana plantations were totally destroyed as was the railroad that had once transported bananas from Portalon to Quepos. The railroad bridge over the Savegre River was gone. The banana company decided not to replant bananas, and in early 1956 hundreds of workers, including Alejandro, were dismissed. He had left Río Sequito seven years earlier and assumed that he would be with the company for many years to come, but suddenly his life was turned upside down. All because of a hurricane called Janet.

Alejandro thought about his situation and realized that things weren't as bad as they first seemed. He was much better off than he was seven years earlier when he left Río Sequito. He had a loving wife and three wonderful daughters; he was in excellent health, still young enough for hard work, and he had some savings.

A friend offered to sell Alejandro a farm near Matapalo. The farm had a thatch-roofed house with a dirt floor, two milk cows, and a horse. There was some pasture, but most of the farm was in secondary and primary forest. For a little over $800 he bought the 23-hectare farm with the cows and the horse with saddle and bridle included. Everything they owned fit into the back of a small pickup that Alejandro hired to move them to Matapalo. Everything was unloaded at the edge of the road, and a neighbor with oxen and a cart hauled it the rest of the way to their new home. Alejandro worked hard, felled trees, cleared brush, and planted pasture and crops. He built a house where Dominga gave birth to four more girls and two boys. All of their nine children were born at home.

Dominga passed away in 1979. Alejandro kept the farm for 13 more years before selling it in 1992 at age 72. He bought a large lot near Matapalo and built houses for himself and several of his daughters. As of this writing Don Alejandro is 98 years old, 62 of which he has lived in Matapalo. He has 12 children, 26 grandchildren, 30 great-grandchildren, and 5 great-great-grandchildren. Still as spry as ever, his hearing is excellent, he can see just fine without eyeglasses, and takes only one pill a day, for blood pressure. He attributes his health and longevity to clean living and hard work. "I'm not a religious man," Don Alejandro told me, "But I talk with God every night. I never ask for anything, but thank him for my health, my good life, and my family."

Don Alejandro

26. Smarter Than We Think

A Glimpse of Intelligence and Adaptability in Tropical Nature

The Jungle Chicken

The great tinamou, called *gallina de monte* or jungle chicken in Spanish, looks like a grouse. With great effort it can fly, but not far or high. When you are walking through the jungle and a horrible ruckus that scares you out of your wits erupts from the base of nearby tree, it is almost certainly a tinamou. The interesting thing about these chicken-sized, gray-colored birds is that they don't always startle easily.

At Hacienda Barú National Wildlife Refuge we discovered years ago that we can use them as a barometer to indicate the prevalence of poachers in the reserve. When lots of illegal hunting is going on, the tinamous are very skittish and will take to flight at the slightest provocation. Although the poachers are usually after paca or peccary, they will kill the great tinamou as well. On the other hand, after several months without being shot at, they become very tame. I have seen them walking along no more than three or four meters away from people who are hiking through the jungle. What amazes me about this is that the change in their behavior takes place within a short period of time. They apparently learn quickly when people are a threat to them.

Sloth Gossip

I have always been fascinated by the extent to which wild animals can learn and adapt to new situations and opportunities. In *Monkeys Are Made of Chocolate* I told the story of a mother sloth who discovered a new way of weaning her offspring.

Pochote trees are native to Costa Rica, but not to the southern zone. In 1987 we planted a copse of pochote in an old pasture next to a primary forest. Pochote trees have lots of sharp spines on the trunks and limbs. Sloths love to eat pochote leaves but must move slowly and carefully through the crowns to avoid the spines.

Weaning an offspring is always a problem for a mother sloth. Since junior doesn't want to leave mom, she must race, at a sloth's pace, through the tree tops and put enough distance between herself and the baby so that the youngster loses her scent and ceases to follow. This is hard work. One mother sloth discovered an easier way. She carried her weaning-aged young into the crown of a pochote tree and left it there. Then she moved completely out of the area. The young sloth, not being as agile as the adult female, took several days to maneuver its way through the spiny limbs until arriving back at the primary forest where it had last been with mom. By that time not even a trace of mother's scent remained, and the young sloth had to face the cruel world on its own.

The amazing thing about this story is that the following year three more mother sloths had learned the same trick. I don't know how they learned, but it is unlikely that they each learned it independently or that they all observed the female that first discovered the technique. There must have been some communication between them. Sloth gossip, I suppose.

Pigs in a Tunnel

When the Highway Department was in the initial stages of the project to construct the final leg of the coastal highway between Quepos and Dominical, we started thinking about the impact that the fast-moving traffic would have on the wildlife at Hacienda Barú National Wildlife Refuge. The project managers from the transport ministry were also concerned, and together we began looking for a solution.

We came to an agreement that there would be thirteen square tunnels specifically for wildlife under the highway and four bridges over it. As eight square tunnels had already been included in the plans for the highway in places where large quantities of water flowed during torrential downpours, we ended up with 21 square or rectangular tunnels, all of which could be used by wildlife when they were dry. For some reason wild animals shy away from round tunnels but don't have a problem with square or rectangular ones.

Both the highway and the tunnels were a new element in the lives of the wildlife, and I must admit that I had my doubts if any animals would use them. To our amazement collared peccaries, which somewhat resemble pigs, began walking through these underpasses even before the construction was completed. We were able to determine this by the peccary tracks entering and leaving the tunnels.

About six months after completion of the highway I set up a trail camera in one of the tunnels to find out what other species were using them. In the first five days the camera recorded three different species crossing the highway by way of the tunnel: collared peccary, coatis and pacas. Large groups of peccary, 20 to 30 individuals, would walk through several times a week (photo below). Over the next six months, in addition to those species mentioned above, the camera recorded opossums, agoutis, skunks, raccoons, capuchin monkeys, anteaters, armadillos, tyras, jaguarundis, and ocelots. We have found puma tracks in a couple of the tunnels but have yet to catch one on camera.

One time a camera captured a video of a female peccary who had recently given birth and whose offspring, about the size of a small house cat, was awkwardly following her through the tunnel. Unlike domestic pigs, baby peccaries are precocial; that is, they are capable of standing and walking within minutes of being born. This particular baby was struggling to keep up with mom, who was tagging along behind a larger group. Not only was this fascinating to watch, but it brought to light an interesting reality. For this baby peccary, only a few hours old, tunnels are part of the natural environment, and the same is true for all peccaries born after the construction of the tunnels in 2010. In one generation the behavior of crossing under the highway has become completely ingrained in the minds of the collared peccaries at Hacienda Barú. It is no longer something recently learned, rather it is just part of life.

The Great Curassow

Called *pavón*, meaning big turkey in Spanish, the great curassow is the third largest bird in Costa Rica behind the jabiru stork and the harpy eagle. In 1999 a study was done of the area covered by the Path of the Tapir Biological Corridor to determine if there was enough biodiversity to merit the expenditure of large sums of money and human effort that would be necessary to carry the project to completion. A dozen different biologists, each a specialist, surveyed the corridor and reported on the diversity of species within their own field of expertise. This included everything from fish and insects to amphibians and mammals. One of the most renowned ornithologists in the country carried out the study of birds.

Upon completion of the study the biologists met with local people who lived in the area covered by the biological corridor. Some biologists listened carefully to the local people and evaluated the information gleaned from them. The ornithologist, on the other hand, totally disregarded any information emanating from local sources. Try as we might, we couldn't convince him that there were great curassows in this region. He didn't see one during his survey, so in his mind they didn't exist here.

Fortunately, the biologist in charge of the study was more inclined to accept local knowledge. Not only was the *pavón* included in the list, it was designated as an indicator species. This is a species whose presence and estimated abundance serves as an indicator of the overall biological health of the area. It was named an indicator species because it is supposedly found exclusively in primary forest, and it tends to avoid all encounters with humans. At Hacienda Barú National Wildlife Refuge, the hilly portion of the reserve is covered with primary forest and the lowlands with secondary forest. So according to the above criteria the great curassow should not be found in the lowlands of Hacienda Barú.

Within a year, to everyone's surprise, visitors, guides, and park rangers began reporting sightings in the lowlands. Today we estimate that there are a dozen mated pairs within the secondary forests of these lowlands. Not only that, but they have become so tame that they pay little attention of visitors on the trails.

Anyone who lives near a rainforest or spends time hiking through one has surely witnessed similar examples of learned behavior in wild animals. More years ago than I like to remember, a high school science teacher told the class that all animal behavior was governed by instinct, and that wild animals don't learn from their environment. Today few scientists or naturalists still hold that erroneous view.

27. Strange Behavior

A Gang of Teenage Monkeys

I once saw a show on the Discovery Channel about a game reserve in South Africa where all of the mature bull elephants had been killed by poachers and a group of adolescent bulls were wreaking havoc, harassing, attacking, and even killing other animals in the park. They were much like a teenage gang. Finally, the owners of the reserve brought in some older, more mature bulls from another reserve, and the young elephants settled down. This situation comes to mind when I think about a strange experience I once had with a group of young white-fronted capuchin monkeys.

They looked like a bunch of teenagers goofing around, teasing each other, and just hanging out. There were six of them altogether. Most were up in the trees relaxing on branches with arms and legs dangling over the sides, or sitting, munching on something. A couple of them would occasionally go to the ground, venture out a few meters from the base of the tree, and quickly scamper back up. None of the monkeys were very large bodied, and from their demeanor I imagined that they were a group of adolescents off by themselves, temporarily separated from the adults in the main troop. This was something I had never seen or heard of before.

Then something got their attention. Leisure activity ceased, and all looked in the same direction. After a minute a big sow raccoon with two small cubs in tow came into view, sniffing around and scratching in leaf litter on the forest floor. The monkeys watched them for a minute, chattered among themselves, descended to the ground, gathered at the base of a large strangler fig and again appeared to be in some sort of discussion. They lined up, side by side, standing upright. Somewhat hesitantly this wall of primates moved toward the mother raccoon. At first, she ignored them, but at some point, the monkeys crossed an invisible red line, and her behavior changed from measured disinterest to full attention. Still not alarmed, she checked to make sure that the cubs were behind her, turned to face the line of monkeys, raised up slightly on her haunches, bared her teeth and hissed at them. I could almost imagine her saying to them: "You want to try it, boys? Come right ahead! Who's first?"

But none stepped forward. The whole gang of monkeys turned tail and ran for the fig tree, scuffled for position, and ascended to the crown. Once safe from the mother raccoon they seemed to lose interest in her, like a person trying to forget an embarrassing incident and pretending it never happened. The monkeys returned to their vagrant behavior, and mama continued about her business with her cubs at her side.

Neither I nor any of the Hacienda Barú guides or park guards have witnessed another incident similar to this one. Mother Nature is constantly coming up with surprises.

28. Torture in the Rainforest

Cats are Cats, Wild or Tame

All of you cat lovers out there aren't going to be very happy with my use of the word "torture" to describe what your beloved pets do to little animals like mice, frogs, birds, and geckos. It's hard to think badly about your beautiful, soft, cuddly friend and easy to rationalize any behavior no matter how abhorrent. The justification usually comes out something like: "Oh! That's just what cats do. She's only playing with the mouse." Humm, I see. Only playing. I wonder what the mouse would have to say about that euphemism.

Diane and I live in a rural environment, and our three cats roam about freely inside and outside. They all love to hunt, seldom eat what they kill, and often don't kill what they catch. I have watched a cat grab a battered mouse in its mouth, toss it into the air, bat it around with its paws, scratch it, and, when the prey quits moving, leave it to die a slow and painful death. You can call that whatever you want, but I call it torture.

I love our pets, but I love nature even more. Though I never really thought about it, it never occurred to me that this type of behavior might exist in the wild. Somehow wild animals, even cats, seemed too noble to engage in anything as abominable as torture.

An excellent field guide on tropical mammals describes how the specialized dentition and formation of the jaw of our wild cats is adapted to killing their prey quickly and efficiently. He tells us: "They kill with a bone-splintering bite to their prey's head or neck, for which the canine teeth—the broadest and strongest of any carnivore—are essential… A gap between the canines and the cheek teeth allows the canines to sink deep into the prey."

Many other sources refer to the death bite inflicted by the large felines. I'm sure that I've heard about it on *Animal Planet* more than once. Cats are almost exclusively carnivorous, seldom touching any food other than animal flesh. In Costa Rica pumas and jaguars eat deer, peccary, pacas, armadillos, rabbits, agoutis, opossums, porcupines, spiny rats, iguanas, bats, snakes, coatis, raccoons, and others.

At Hacienda Barú National Wildlife Refuge we have nine trail cameras which are located in places frequented by wildlife of all kinds. Seldom does a month pass when we don't record at least one puma, the largest carnivore found on the refuge. Usually, the photo or video just shows them walking past the camera, but on April 19, 2017 one of the cameras captured a series of 108 photos of a female puma and her prey, over a period of 32 minutes. The prey is definitely an opossum, but the species is unclear as the coloration appears to differ in various images.

After studying several photos and consulting with our guides, I think it was probably a water opossum, sometimes called a yapok. The location is right next to a stream, a likely place for that species to be found. In the images it appears like the immediate area was lit up, but the lighting is infrared and not visible to the puma or the opossum. Infrared rays emitted by any warm-blooded animal will activate the camera. This camera was set to take a series of three photos, pause for five seconds, activate again, and take three more. I want to share some of those photos with you. They completely shattered my idealistic conception of the nobility of wild cats.

At 03:09 a.m., an adult puma arrives at the location and activates the camera. At 03:11 her head appears in the lower left corner of the photo. She seems to be looking at something on the ground.

Thirty-seven seconds later at 03:11:57 we catch our first glimpse of the opossum who appears to have already suffered enough physical abuse so that it is unable to escape. In the image it appears dead, but in the series of three photos it is obvious that it can still walk.

03:13 – She is looking at the opossum on the ground. It is hidden behind her head.

03:14 – They have moved back toward the stream.

03:16 – The puma and the opossum have moved several meters up the slope, and she is batting it around with her paw.

03:22 – They are back down near the stream again, and she has the prey in her mouth. Next photo shows she has carried it back up the slope and is batting it around again. The opossum is difficult to see in this image.

03:23 – Only 12 seconds later they are back down by the stream. Again, only the eye of the prey is visible.

03:39 – Almost 17 minutes later the opossum is still alive and moving. The movement is only visible when scrolling through the series of three photos.

03:41 – The puma leaves the scene 32 minutes after arriving.

In pondering this situation and looking for an explanation that is consistent with the typical description of a "bone splintering bite to the head or neck," I believe that this puma would've acted much differently if the prey had been larger and able to fight back. The tusks of a peccary or the antlers of a deer could definitely do some damage, and the puma would probably go for the fastest kill possible. Even the claws or teeth of a coati, raccoon, or paca would give a puma second thoughts. But this opossum was totally defenseless, allowing the large cat a chance to "play" with its prey. Mother Nature often acts in strange ways, and the more I delve into her mysteries the more I realize how little I really know of her secrets.

03:16:12 The puma is playing with the opposum.

03:23:06 Puma and prey are back down by the stream.

03:39:50 Almost 17 minutes later the opossum is still alive and moving. The movement is only visible when scrolling through the series of three photos.

03:41:09 The puma is leaving the scene, 32 minutes after arriving.

29. Big, Beautiful, Left-Handed Birds

The Return of the Scarlet Macaws

Once upon a time there was a Bribri Indian chief named Pabru Presberi who was master of some big, exceptionally beautiful birds called "Pa." Everywhere the chief went, large groups of Pa flocked around him. Some were red and others green, and all were covered with stunning, bright-colored plumage. One day some strange men with beards from a faraway place called Spain arrived on the shores of Costa Rica. Awed by the beauty of the Pa, the Spaniards killed them for their feathers which they took to their homeland as gifts for the royalty.

The word "la" in Spanish means "the" in English, so they referred to the birds as "la pa," which overtime became lapa. As time went on there were fewer lapas. The Spaniards had killed all but the large flock that followed chief Pabru everywhere. Eventually the chief led a revolution against the Spaniards, so they captured him and took him to Cartago where he was imprisoned.

The lapas followed. Eventually the Spaniards executed the chief and the red lapas flew away to the Pacific coast and the green lapas to the Caribbean coast. To this day that is where they reside.

Of course, in English, the *lapa roja* (*Ara macao*) is the scarlet macaw, and the *lapa verde* (*Ara ambigua*) is the great green macaw. Their numbers have diminished drastically over the years to the point that the great green macaw is now listed on the IUCN Red List as endangered. The scarlet macaws, which were once seen in large colorful flocks up and down the entire Pacific coast, have also diminished, but not to the extent of their green cousins. It is estimated that today there are about 600 scarlet macaws in the Osa Peninsula and 400 in Carara National Park. They disappeared from the area between Manuel Antonio and Uvita in the 1960s.

It is the same old story that has led to the local extinction of so many species like tapirs, jaguars, white-lipped peccary, and brocket deer, among others, due to habitat loss and excessive hunting. At the beginning of the last century the area around Hacienda Barú was covered with rainforests replete with all of the species mentioned above. As settlers colonized the area, they cleared the land and killed the wildlife. As the forest was felled, the large old trees, so necessary for nesting sites for the scarlet macaws, were lost forever, and reproduction of the scarlet macaws diminished.

Hunters killed the birds for their feathers, and for the pure perverted pleasure of seeing them fall dead. One former resident of Dominicalito told me of an occasion when he was talking to a man who had been out hunting and had a 22-calibre rifle with him. It was late afternoon, and seven scarlet macaws flew into a nearby tree to roost for the night. "Watch this," exclaimed the smiling hunter. "Those birds are so stupid that they will just sit there and let me kill them all." One by one he shot them off of the branch and watched them fall to the ground. "See what I told you?" he bragged. "Look how stupid they are."

In the mid-1960s there were five scarlet macaws left in the village of Hatillo, and two local hunters vowed to kill them all. They made a bet of ¢100 colones (about $15 at the time). Each claimed that he would kill more of the five. After a few days each hunter had killed two scarlet macaws, but the remaining bird had evaded them both. One day the two were talking when that last big red bird flew into a tree within shooting distance. Both took aim and fired at the same moment, and the bird fell dead. But neither could claim the ¢100 colones as there was no way of knowing which hunter's bullet had killed the last scarlet macaw in Hatillo.

But times change and people change. By 1985 Hacienda Barú, with 180 hectares of rainforest, was the only forest reserve of any size left near Dominical. All the surrounding area, with the exception of a few small parcels of forest on the steepest hillsides, had been denuded. That was the peak of the deforestation and marked the beginning of the restoration of natural vegetation to the area.

Land usage has changed from farming and ranching to recreation and tourism, bringing with it the regeneration of much of the forest habitat where wildlife once abounded. And the wildlife is returning. When I first came to Hacienda Barú in 1972 the white-faced capuchins were the only species of monkey. Since that time, spider monkeys, howler monkeys, and squirrel monkeys have returned. Their numbers are still small, but they are growing. The puma has returned, a strong indication that the ecological health of the area is in good enough condition to support large predators.

At the 2009 annual membership meeting of the Asociaciòn Amigos de la Naturaleza (ASANA), Don Walter Odio, a member of the board of directors of ASANA and owner of Rancho La Merced Nature Reserve in Uvita, interrupted the meeting to announce that he had just received a text message that a group of scarlet macaws had landed in some trees at Rancho La Merced. It was the first sighting in Uvita in over 50 years.

That same year some macaws were sighted in the Dominical area and at Hacienda Barú. In 2012 scarlet macaws acted like they wanted to nest in a large ceiba tree, but it was a young tree, and they probably couldn't find an adequate hollow that would work for a nest. This prompted us to build some nesting boxes. We secured four of them in the tops of trees where we had seen the macaws, but the big red birds didn't seem interested. One day a Hacienda Barú guide thought he saw some movement near the opening of one box. Thinking a chick might be inside, he focused his spotting scope on the opening only to discover that the box was inhabited by a hive of Africanized bees. We checked the other three boxes, and all were full of bees. Getting the boxes out of the trees was quite a chore, but we managed it, and don't plan on putting up any more nesting boxes until someone figures out how to make them bee proof.

The most critical threats to the macaws are the poaching of chicks from nests to sell on the black market, and lack of adequate nesting trees. Hunting no longer appears to be a major problem, and over the last 25 years the poaching of chicks has become less of a threat than it used to be. The reason for this is that the demand for scarlet macaws for pets can now easily be supplied by chicks raised in captivity. Wild macaws poached from their nests no longer bring the high prices they once did. Lack of adequate nesting sites is still the main threat to the big red birds. Fortunately, macaws have been observed nesting in some very fast-growing softwood trees called gallinazo and balsa which are common in newly rewilded secondary forest. Hopefully, the macaws will learn to nest in this type of tree.

The scarlet macaws that have been seen in Dominical and Uvita have probably migrated into the area from the Osa Peninsula, and the ones in the area around Quepos have probably come mostly from releases of birds raised in captivity. The first release took place in 1998 when the former owner of Gaia Rescue Center released 21 birds at El Silencio near the Savegre River. These macaws were a mixture of birds raised in captivity at Gaia and birds that had been brought to the rescue center by wildlife officials. Most had been confiscated from people and hotels who were holding them in captivity illegally. The year after the El Silencio release, six pairs of scarlet macaws invaded houses, offices, and tool sheds within the community and made their nests in mattresses, cupboards, closets, and attics. They made quite a mess. Some of them successfully reared chicks. The second year they all ventured off into the wild.

Ana Marìa Torres, MsC, is executive director of the non-governmental organization ASOMACAO, which now operates an aviary in the same location as the old Gaia rescue center. Among other things they raise scarlet macaws and release them into the wild. As of this writing, they have 24 birds, 15 of which are for reproduction and 9 for release. Additionally, they have secured nesting boxes in trees where semi-wild birds can raise their chicks. From one of the nesting boxes seven chicks have been successfully raised and fledged in three years.

Ana María, a veterinarian, who holds a master's degree in Conservation Medicine, is totally dedicated to the ASOMACAO project. When she asked Hacienda Barú for help climbing into the treetops to place the nesting boxes, we readily agreed. She has learned to climb and now attends to the nesting boxes herself. She once placed a camera near the opening of a nesting box to watch for activity within. One of the chicks stood in front of the box opening constantly for about two weeks waiting, for the adults to tell it the time to fledge had arrived. One can only imagine it standing there whining, "Mommy, mommy, when can I go? When can I go? I want to go now."

Punta Leona Hotel is located near Carara National Park, and many scarlet macaws hang around the location. The hotel staff have been observing macaws for a long time and know their habits well. On their website there is a list of ten curiosities about the big red birds. Number 7 really caught my attention. It says that most scarlet macaws are left-handed. They use their left legs almost exclusively to eat and groom or whenever they need to grasp something. Ana María is going to pay close attention to the macaws at ASOMACAO to see if they too are left-handed.

30. Collared Peccary

Intelligence to Rival That of Monkeys

Although I have lived at Hacienda Barú since 1972, I had never seen a peccary in the wild until 1997. Volunteers had come here, stayed a week or two, helped on environmental projects, and during that short time many of them had the good luck of seeing the collared peccary (*Pecari tajacu*), called sainos locally. Our guides had seen them on numerous occasions as had our guests.

My brother Rex came for his first visit in many years. He and his wife LaVonne did an overnight jungle camping tour with a sharp-eyed local guide named Deiner. I volunteered to carry the food up to the camp that afternoon. There had been a lot of collared peccary sightings recently, and I figured I had a good chance of seeing one during my hike. I saw tracks everywhere and even caught a whiff of their musky scent. The ground had been dug up in several places where they had rooted around for invertebrates, roots and tubers. When I got to the camp with the food, Rex and LaVonne couldn't wait to tell me about the group they had seen.

"They were just a little way off the trail," said Rex excitedly, "a big, dark gray female and a couple of young. The babies were a reddish color."

"I don't believe this. It's unreal!" I exclaimed in exasperation. "I've been looking for a collared peccary for 25 years, and you see them the first time you go for a walk in this jungle. It's not fair."

"It was pretty cool," Rex went on, oblivious to my ranting. "The female gnashed her teeth together till they clattered and kind of shook her head at us, but the little ones seemed curious. Deiner said not to worry. He had a big stick in his hand just in case they did something unexpected. The female stood about as high as my knee and had her lips curled, baring her tusks. She could do some damage if she took a notion to." Rex went on to say that after a few minutes the big female led the group of sainos off into the jungle. The two piglets followed reluctantly.

In my quest to find the collared peccary at Hacienda Barú National Wildlife Refuge, I often smelled them. Sainos have a scent gland on their rump about a finger's length above the tail. Rubbing the scent gland on a tree trunk and defecating and urinating at the base is a method of marking territory. The smell is strong and musky. Another common sign of the presence of peccary are their small, sharp-pointed tracks in the mud. These are similar to deer tracks but are smaller and tend to point downward.

On several occasions I was close enough to hear the characteristic alarm bark emitted by the collared peccary when excited. Once I surprised a big one that let out an enormous "hurummpf" that nearly scared me out of my wits. It then went charging off through the jungle, "hummpf, hummpf, hummpf." All I got to see of that one was the movement of the foliage as the saino forced its way through the low-lying vegetation.

When it finally came, my first sighting was no big deal. Three collared peccaries ran across the trail in front of me and off into the rainforest. I only caught a glimpse of them, but I had officially seen them. From that day forward sightings have become much more frequent. Populations have increased due to diminished levels of hunting, and the peccary have become a little less spooky due to regular contact with nature lovers hiking on the trails. They still keep their distance, but don't panic at the sight of humans.

To me the most amazing thing about peccary is that they still exist. Since the first humans appeared in the Americas more than 10,000 years ago, wild pig-like mammals have been hunted for their meat and hides. Yet both Central American species and a South American species have managed to survive in reasonably large numbers, while many other heavily hunted mammal species have become extinct or nearly so.

Both Costa Rican species, the collared and the white-lipped peccary, are considered threatened and are listed in appendix II of the CITES convention for the control of trade in endangered species. Large herds of the latter once wandered the wilds of Corcovado National Park. Reduced park budgets and fewer ranger staff have resulted in an uncontrolled wave of poaching which has severely

threatened the white-lipped peccary and other species. In the area around Hacienda Barú National Wildlife Refuge, however, collared peccary numbers are higher than ever. This is due mostly to increased vigilance by private property owners.

In spite of the threat from human predation, peccaries range from Arizona to Argentina in every type of habitat imaginable, from dry desert to extremely wet tropical rain forest. In fact, their adaptability has probably been their most powerful survival tool. In the deserts of Mexico and the southern United States, prickly pear is the collared peccary's preferred food. In the humid Central American tropics, roots and tubers form the major portion of their diet. Palm nuts and other large seeds are also a favorite menu item. All peccaries will eat meat but aren't very well equipped to hunt and capture wild game. The protein in their diet consists mostly of large insects, snails, and carrion.

Intelligence has also helped the peccary to survive. In 2010 when a new highway passed right through the middle of Hacienda Barú National Wildlife Refuge, we convinced the highway department to build tunnels under the road to facilitate animal crossings and diminish roadkill. The collared peccary was the first species to use these subterranean crossings. We found their tracks entering and exiting the tunnels even before the construction was finished. Considerable roadkill of other species still occurred for about six months after the opening of the highway.

Eventually most mammal species learned to cross under the traffic rather than through it and thus avoid death on the asphalt. With trail cameras we have documented 13 different species that cross through the tunnels, but the peccary were the first. I have never seen a dead peccary on the road. On the other hand, it took the white-fronted capuchin monkeys over a year to start using the bridges we built over the highway. During that year, about half a dozen monkeys died crossing the highway on foot.

Other than humans, peccaries are preyed upon only by Costa Rica's two large felines, the jaguar and the puma. At Hacienda Barú the puma is their only serious natural predator. I suppose it is possible for a hungry ocelot to kill a straggling baby, but babies seldom stray from their mother's side, and I doubt if an ocelot would want to risk an encounter with a furious female's tusks.

It appears that in the area around Hacienda Barú there is an equilibrium between peccary numbers and predation. The population is high and appears to be stable. There are enough collared peccary here that I now see them regularly. There is better than a 50-50 chance that a visitor who walks through the Hacienda Barú National Wildlife Refuge on one of the trails for a couple of hours will see one.

31. The First Gringo in Dominical

The Amazing Life of Thomas J. Brower

When Thomas J. Brower first came to Costa Rica during World War II, the US was worried about the Panama Canal. If it were to fall into enemy hands, Allied naval capacity would be severely crippled. A land route, imperative for the defense of the canal, did not exist. Building one became top priority, and the Pan American highway was the result. The US government contracted with the Charles E. Mills Construction Company to build the 300 km stretch of road from Cartago to the Panamanian border.

Tommy Brower landed in Dominicalito with the construction company in 1940. A crude dock was built and heavy machinery unloaded. Part of the crew, with the help of local labor, built a camp for the company workers, and the others started building a road to San Isidro. Once the trocha, as the rough-cut road was called, reached San Isidro the workers split into two groups with one working toward Palmar Norte and the other toward Cartago where they would eventually meet up with other crews working from those locations toward San Isidro. It was a major operation, the magnitude of which had never been seen in Costa Rica.

Thomas Brower began as a bulldozer operator and was later promoted to supervisor. As such he was stationed in San Isidro, which was to become his home for the next ten years. It was there that he met and married a lovely young lady named Consuelo Aguerro, ten years his junior. For the first decade of their marriage the couple lived in San Isidro. Five of their seven children, four girls and one boy, were born during that decade. The last two boys were born after the family had moved to Dominical. Regardless of where they were living at the time, doña Consuelo always went to San Jose to bring her new bundle of joy into this world. In those days it was common for women to give birth at home with the assistance of a midwife, but Tommy wasn't about to risk losing the love of his life during childbirth for lack of qualified medical attention and a modern hospital.

Once his work with the construction company was finished, Mister Tommy, as he came to be called affectionately by all who knew him, became involved in other business ventures. In San Isidro he and Consuelo were especially remembered for all the wonderful things they did for the community. Their second daughter Sara Isabel remembers the Christmas parties they sponsored for the poorer children in the area who weren't likely to receive any gifts. The couple was also a driving force in the founding of the Red Cross in San Isidro, and donated the lot where the original headquarters was built. They were considered to be important community leaders.

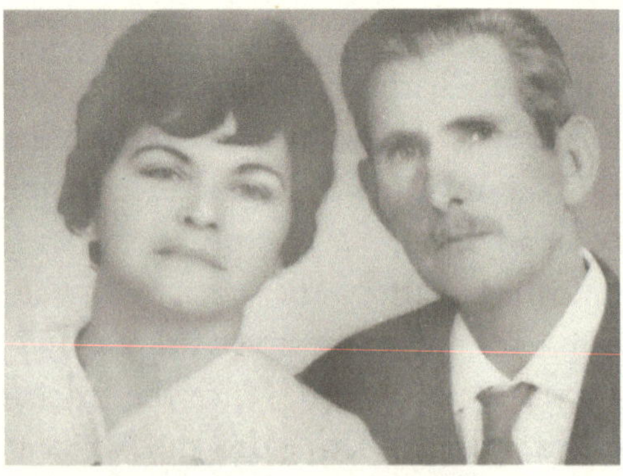

"Mister Tommy" Brower and his wife Consuelo

The Brower family was living in San Isidro in 1948 when the revolutionary troops of Don "Pepe" Figueres arrived to claim the city. In anticipation of the impending battle, the head of the municipal police came to their home and draped an American flag on the front of the house so that combatants on both sides would know that the inhabitants had nothing to do with the civil war. On March 16, 1948, government troops under the command of General Tijerino arrived by sea and landed in Dominicalito. They marched to San Isidro where they clashed with the revolutionaries in a fierce battle that lasted 36 hours. General Tijerino was defeated and killed. A victorous Don Pepe returned to San Jose, which he captured, and officially ended the war.

On April 20 a second detachment of troops still loyal to the now deposed and exiled government, landed in Dominicalito and marched to San Isidro. They dug in on the hill where today we find the installations of the Ministry of Transport. Either this detachment didn't know the civil war was over, or just didn't want to give up, but when don Pepe's troops returned to San Isidro, what ensued was more of a slaughter than a battle. There were so many casualties that the community had to build a make-shift crematorium to dispose of the corpses. Fortunately, the combatants on both sides respected the American flag on the Brower home, and no one in the family was harmed, nor was their property damaged.

In 1952 the Browers moved to Dominical and built the Dominical Hotel, a beautiful two-story structure, on the lot where today we find Restaurant El Coco. The venture was a resounding success. People came from as far away as Cartago to stay at the hotel. Although they had to ford a river and walk the last three kilometers to get there, it became a weekend party destination for the people of San Isidro. Two of the children, Sarah Isabele and Thomas Jefferson, were too young to join the parties, but they remember peeking through the cracks to watch the adult festivities.

Later Mr. Tommy, in partnership with Misael Ceciliano, built the Miramar Hotel and Restaurant where today stands Roca Verde.

Once every two weeks a pilot named "Sapi" brought in food and liquor for the hotels in his small plane. A lady named doña Angela oversaw a telegraph that connected Dominical with the outside world, and she took care of sending the orders to Sapi. Coco, a fisherman, brought the fish. Tommy loved to hunt for oysters on the rocks between Dominical and Dominicalito which are now known as Rocas de Amancio. The rainforest provided meat for the restaurants. Tommy and Consuelo used to accompany Manuel Angel Sanchez on his hunting excursions into the jungles of Hacienda Barú, where he was the foreman. Their quarry was mostly collared peccary and paca, but great tinamou and great curassow also adorned the plates of the guests at times.

Among her other talents doña Consuelo was a midwife who assisted many women during childbirth. Half a dozen of doña Blanca Valverde's children were delivered by doña Consuelo. The year 1955 brought the worst flooding in the history of Costa Rica. When the torrential rains hit the area, doña Blanca was seven months pregnant, was suffering from a relapse of malaria and had a nasty infection in her foot. The water rose so high that the entire town of Dominical had to take refuge on the second story of the hotel. After six days of pouring rain the storm ended, the water level dropped, and a Red Cross boat was able to reach Dominical. Mister Tommy and other neighbors helped Manuel Sanchez, doña Blanca's husband, carry her to the boat. Doña Consuelo accompanied her to Puntarenas and from there on to San Jose to the San Juan de Dios hospital by bus. After three weeks in the hospital and three weeks with an uncle in Escazú, doña Blanca, by then eight and a half months pregnant, managed to return to Dominical—bus to San Isidro and small plane to Dominical—where two weeks later, she was assisted by doña Consuelo in giving birth to a healthy son.

The Browers were active participants along, with Rafael Angel Aguero and Marina Badilla, in the establishment of the first school in Dominical where Doña Marina was the first teacher. Tommy and

Consuelo's children attended the school together with the rest of the children in town.

In 1968 the Brower family moved to San Jose for a year and then to Stanton, California. Today the oldest daughter, Hilda, lives in Norway. Sarah Isabele in San Jose, Costa Rica, Donald Francis in San Vito, Costa Rica. Jeannette in South Carolina, and Thomas Jefferson and John Carl in California. I owe granddaughter Kelly from Uvita a big thank you for introducing me to the rest of the family whom I had the great pleasure of meeting in November of 2018. I also want to thank another granddaughter Yessenia for the wonderful photos she scanned from the family album. And of course, I am thankful to the entire family for all of the information they supplied to help document this history.

32. Machetes

An Absolute Necessity of Life to Campesinos

Many years ago on a Saturday afternoon, having gone to the police station in Matapalo for something or other, I found a hand-cuffed, bare-footed, blood-covered man with a deep cut starting on his right shoulder and extending across his chest. There was blood on the floor, the chair he was sitting in, and even on the wall. Don Marcos, the policeman, was sitting at his desk writing up a report. "Machete fight," he said in reply to my inquisitive look. "The other one's on his way to the hospital. This one started it," nodding toward the bloody prisoner.

To the average outsider visiting rural Costa Rica the big knife is simply a machete, but to the campesino, or rural resident, it is an absolute necessity of life. Machetes are used for everything: chopping weeds, cutting small trees, trimming hedges and bushes, pruning trees, peeling the bark off poles, splitting kindling, peeling oranges, harvesting rice, corn, cacao, bananas and other crops, cutting boards, shaving the edge of a board, scraping crud off of anything, unscrewing bolts, cutting lawns, digging in flower beds, killing snakes…I'm sure there are more uses that can be added to the list. I have never seen anyone shave with one, but I don't doubt that has happened. As mentioned above, it can be used as a weapon.

I wouldn't be surprised if Costa Rican campesinos have as many names for the long-bladed knives they use in their daily lives as Eskimos have for snow. Not all campesinos from all locations will agree with these names as they vary from one place to the next.

Macheta – an all-purpose machete that is usually carried in a scabbard hung from a belt

Guapota – a large machete between 24 and 28 inches long that is used to cut down thick secondary growth. It is much wider near the point than the haft.

Cutacha – a shorter version of the guapota, between 20 and 24 inches long, often used in palm plantations to hack through the fibrous leaf stems.

Rula – a narrow blade, 26 to 28 inches long, used to chop weeds in pastures. It is very efficient for cutting non-woody stems.

Machete pala or machete de suelo – a short, very wide-bladed machete used to cut lawns and as a trowel in gardening. It works great for putting a nice edge on a lawn.

Machete curvo – a medium-size machete that is curved or doubled over at a 45-degree angle in the middle. The part near the haft is narrow and the part after the bend is wide. It is used like a scythe to harvest rice, corn, and other grains.

Cruzeta – a long, narrow machete similar to a sword that years ago men wore for a night on the town. The blade is narrow and is distinguished by a cross-guard separating the blade from the haft. The idea is that in a barroom fight the cross piece will protect the wielder's hand from an opponent's blade.

Campesinos carry a special file, a lima, to sharpen their machetes. It is a flat metal file about a foot long with one edge ground to the sharpness of a knife. With my own machete I just use the file like I would on a shovel, and it works fine, but everybody tells me that I'm doing it wrong. I must admit that my worker's machetes are sharper and hold an edge longer than mine.

Machetes became quite popular after the release of the first of a series of films starring Machete, a mean-looking character played by Danny Trejo. As you might suspect his favorite weapon is the machete, and from the scars on his face, most of which I suspect, are the creations of make-up artists, you might imagine that he has been in more than a few skirmishes. The machetes he holds in most of his photos are guapotas.

Back in the days when the Osa Peninsula was so wild the police wouldn't even go there, my good friend Patrick made his living walking over the entirety of what is now Corcovado National Park buying gold. Pat's mother-in-law owned a pulpería or general store where they also served liquor. After a few too many drinks hot-headed miners were often inclined to fight each other with any weapon at hand, and machetes were usually close by.

"There are three kinds of machete fights," says Pat, "the first is when both men have machetes, the second when one has a machete and the other doesn't, and the third when one has a machete and the other a gun. The first type never happens. Both fighters know that even if they win, they're going to get cut, and neither wants that. All it takes to defuse the situation is for someone to talk them out of it. I've prevented lots of bloodshed just by talking to the fighters." Pat goes on to say that in the second and third types of machete fights, the fight always takes place, and the disadvantaged fighter ends up dead.

Many of the men who work at Hacienda Barú were practically born with a machete in their hands. When asked if any of them had ever witnessed a machete fight, only Juan Ramón said he had. "It was more like a dance than a fight," he laughed. Juan went on to say that the two hot heads had left the Rancho Memo bar and moved on to the Dominical soccer field to fight. They circled each other, never getting within range of the other's knife, and occasionally making a jab. Once the fighters saw the local policeman arriving on the scene, both ran away before he could arrest them. This was a type-one fight, and, just like Pat said, neither fighter wanted to get cut.

I only know of one machete fight in this area where one of the participants died. In the 1940s Santana Jiménez, his brother Sen, and a couple of workers were driving a bunch of fat cattle to a barge in Dominicalito to be shipped to market in Puntarenas. They stopped at a cantina along the way for a bite to eat and a couple of drinks. Sen and the workers went outside while Santana paid the bill. There were a couple of shady-looking characters in the cantina who spoke with Nicaraguan accents. As Santana was on his way out the door, one of them made a vulgar remark about his daughter. Santana took offense and in the ensuing fight, was hit straight across the forehead with a machete. The blade penetrated his skull and embedded part of his straw hat in the wound. He received another deep cut on his waist. Sen took up the fight, severely cutting one of the assailants and chasing them off. A few days later word got back that the two Nicaraguans had fled as far as Parrita where the wounded one died. Santana's wound became infected; he died a couple of weeks later.

For many years at Hacienda Barú we have combated trespassing and illegal hunting by capturing the hunter's dogs and taking them to the police or wildlife authorities. In the mid-1970s my workers captured three hunting dogs and took them to the road where I could pick them up. About the time I arrived so did the hunters. One of them was furious, yelling and waving his machete around like a mad man. He said that if I didn't give his dogs back, he was going to cut my head off. I didn't have a machete and the situation was beginning to look like a type-two machete fight. Backing away from him I maneuvered myself alongside the Jeep, until I could reach in and grab my revolver from the glove box.

Seeing that the situation had changed and we were now in a type-three fight, the enraged hunter turned, threw his machete off to the side of the road, and stood there shouting obscenities at me. His brothers ran up, grabbed him, and took him away. I took the dogs to the policeman in Matapalo. That was as close as I have ever been to being in a machete fight, and I never want to be any closer.

33. Bone Breakers

Raptor? Scavenger? A Very Special Bird!

I once posted a photo of a crested caracara on Facebook along with the caption that it is my favorite bird. Among the comments was the question, "How can you have a favorite bird?" Apparently, that means that I should love all birds equally. Maybe so. Nevertheless, there are several that make my heart beat a little faster whenever I catch sight of them, but none so much as the crested caracara. *The Birds of Costa Rica* by Richard Garrigues says that the crested caracara is the national bird of Mexico, so I guess I'm not the only one whose heart goes pitter patter at the sight of one.

Many years ago, when Hacienda Barú was a cattle ranch, we used a tractor and a mower to chop the weeds in the pastures. It was a fun job because the sound of the tractor attracted flocks of cattle egrets and a scattering of other birds. Insects, lizards, frogs, and other small animals ran from the machine. The egrets walked and skipped alongside, fluttered ahead in short bursts, grabbing and eating every living thing small enough to swallow.

Once a large rat dashed out, barely avoiding getting killed by the mower, only to get pecked twice, rapid fire, on top the head by the straight sharp pointed beak of an egret. It lay there writhing and tried to run but could hardly move. Down swooped a roadside hawk going for the kill, but missed the rat, and right behind it came another raptor, twice as big as the hawk. It easily snatched up the rat and carried it off to the edge of the pasture to enjoy a hardy meal.

The bird turned out to be a crested caracara, and it was the first and only time I have seen one behave like a raptor and capture a live prey. Normally they are scavengers and sustain themselves on carrion, the rotting flesh of dead animals. It isn't unusual to see a crested caracara feeding together with a group of vultures. Though much smaller than a black vulture, when a crested caracara arrives at a carcass, the large, black scavengers move over and give it plenty of room. It is a strange sight to see such a beautiful bird pecking away at a carcass while the vultures stand aside and watch.

The Spanish name for the crested caracara is the *quebranta huesos* or bone breaker. Though I have never seen it happen, I know two people who have witnessed one of these birds carry a bone high into the air and letting it fall on a hard surface, like rocks on the beach. If the bone doesn't break on the first try, the caracara repeats the process until it does. Once the bone breaks, the caracara plucks out the nutritious marrow from within.

Beautiful, powerful, dignified, respected by larger scavengers, and thrifty. What is there not to admire about the bone breaker?

34. The Ocelot Factory

Finding New Territory May Involve Crossing Roads

On September 14, 2016, one of the Hacienda Barú guides sent me a photo of an ocelot killed on the coastal highway in front of the Hacienda Barú National Wildlife Refuge. From the spot pattern on its neck and shoulders I was able to identify it as a young male, about two years old. We had named him Frodo. His identification was made possible by comparing the photo of the dead ocelot with photos and videos of Frodo taken by our trail cameras. Every individual ocelot's spot pattern is unique, making identification easier. There are tunnels under the highway at Hacienda Barú, and most of the ocelots use them. Maybe Frodo had never crossed the highway before and didn't know about the tunnels.

A couple of months earlier a new mature male, an enormous individual who I named Brutus, had ousted the male who had reigned for at least three years. Maybe Frodo, who was probably a son of the ousted alpha male and who was approaching reproductive age, felt threatened by Brutus and was forced to leave the territory where he was born and raised. The death of the beautiful, young spotted cat was sad, but not cause for alarm. There were at least one producing adult female and an immature female, Tinkerbell, about a year from reproductive age, and quite possibly a second mature producing female within Brutus' territory. So, the death of Frodo wouldn't have a major impact on the population. We thought.

Within the next year four more ocelots were killed on the highways within five kilometers of Hacienda Barú National Wildlife Refuge. This was definitely cause for concern. Why were so many ocelots, being killed on the roads? In May 2017 a friend sent me a photo of an ocelot killed on the highway near the village of Barú. It appeared to be a young female, but the person who took the photo didn't pay much attention and reported only seeing a dead ocelot. It would have been helpful to know as much as possible about this roadkill, the sex, age, size and physical condition. These simple data could give us clues as to the causes of the elevated level of death on the highways in the area.

The size of an ocelot's territory is extremely variable. According to Mark Wainwright's excellent book, *The Mammals of Costa Rica,* an individual ocelot's range can vary from 100 to 3,100 hectares. I would imagine that the difference has to do with the abundance of prey. Ocelots eat primarily small rodents such as spiny rats, though they will also eat larger rodents such as pacas and agoutis, as well as iguanas, small caimans, coatis, raccoons, young peccary, opossums, and many more. In other words, they will eat about anything that they can catch and kill. It seems logical that there would be a high density of ocelots in a rainforest with a high level of biodiversity and an abundance of prey. A place such as Hacienda Barú National Wildlife Refuge.

Between September 7 and October 10, 2017, the Hacienda Barú trail cameras in the highland rainforest captured photos of four different ocelot individuals: Tinkerbell and her mother, and another unknown individual whose sex we were unable to determine from the photos. During the same period, one of the lowland cameras near the beach captured a photo of a large cub less than a year old. I had seen this cub and a litter mate with their mother in previous months and was reasonably sure that they were all still in the area. If so, that means that our trail cameras captured photos of five different individuals on 330-hectare Hacienda Barú National Wildlife Refuge in one month, and there are quite possibly two more. That is much higher than normal density.

Apparently, the death toll from crossing the highway isn't having much impact on the ocelot population. In fact, the high density of ocelots is very likely one of the primary factors influencing the exit of some individuals from the area thus resulting in an elevated amount of roadkill. Young animals, feeling the pressure from the large number of competitors, are leaving the territories where they were born and raised, and some of them are being killed on the highways. We needn't worry so much that these deaths are negatively affecting the population, rather it is over population that is forcing them to leave the relative safety of the rainforest where they have lived to look for new territories. After seeing all the ocelot photos from September-October I commented jokingly that we must have an ocelot factory at Hacienda Barú.

Much of the above, of course, is pure speculation. You might call it an educated guess. We would prefer to determine the reasons for the large number of ocelot deaths on the highways through scientific research. At the moment Hacienda Baru National Wildlife Refuge is in conversations with the environmental organization Nãi Conservation about working together to study the ocelots in hopes of better understanding their biology and behavior at Hacienda Barú and in the Path of the Tapir Biological Corridor. Nãi is a great group, and we look forward to working with them.

Illustration by Jan Betts

35. Feral Pigs

Why Aren't There Any on Mainland Costa Rica?

In the mid-1980s I had the opportunity to visit Caño Island in a small and none-too-speedy boat. The trip from Drake Bay to the island took more than an hour. Capitan Felix, owner of the boat, was a talkative soul, and fortunately his chatter was not the least bit boring. We learned the story of the island which had, at one time, been plagued with feral pigs.

In 1976 when the government created Caño Island Biological Reserve, the wildlife department decided it was important to rid the reserve of the feral pigs. Capitan Felix was part of a group of four local hunters who had been contracted to hunt down the pigs and kill them. They assumed that hunting feral swine would be like hunting peccary on the mainland, but then they found out that the wildlife department wouldn't let them bring their dogs into the reserve. Their only alternative was to walk the length and breadth of the island repeatedly and shoot every feral hog they saw until none remained. Official estimates were that 40 to 50 of the pigs wandered over the 3.26 square kilometers of Caño Island, and the hunters figured they could probably accomplish their task in a month.

At first the killing went fast, five or six pigs a day, but within a week it slowed considerably, and the hunters were lucky to shoot one or two a day. There were days when they didn't kill any. After a month there were only four feral pigs left, but they were the craftiest and quickest of the lot. Even catching a glimpse of them was difficult. One day the caretaker on the island found Capitan Felix by himself, sitting on a log taking a breather. "Listen, Capi, you didn't hear this from me," he winked, "but next Wednesday both game wardens are taking some days off. They'll leave the island and won't return for eight days. I'll be here alone. Do whatever you want to kill those last four pigs. It's not my job to supervise your work."

Early Thursday morning the Capitan and one companion left in his boat. A couple hours later they were back with four hunting dogs. By dusk Caño Island Biological Reserve was free of feral hogs. I never asked Felix what they did with the dead pigs, but I would guess that they ate their fill of pork, took some choice cuts home and fed the rest to the sharks.

Pigs were probably introduced to both Caño Island and Cocos Island at around the same time. Some sources say that pirates brought pigs and left them on the islands in hopes that they would reproduce and provide food for future visits. British and European explorers also sailed the Pacific during the seventeenth and eighteenth centuries, and they may very well have been the culprits who brought the pigs. The most reliable information comes from a genetic research study that claims the 400 to 500 feral hogs on Cocos Island were descended from one pair of domestic pigs with genes from both Asian and European swine which arrived in 1793.

Regardless of how and when they got to the island, their presence is extremely detrimental to the ecosystem and eliminating them entirely would be next to impossible. Not only does Cocos Island have 10 times as many feral pigs as did Caño Island, it has around 15 times more land area as well. Someone once suggested introducing several male jaguars on Cocos Island in hopes that they would kill all the pigs. Who knows? It might work, but I doubt it.

In the United States, the US Department of Agriculture estimated there were seven million feral pigs in 35 states and that they did about $2.5 billion dollars worth of damage annually, mainly to crops (2014). To eliminate, or at least control, feral swine in certain target areas, $75 million was budgeted.

Genetically these animals are slightly different from the Cocos Island feral pigs, but are closely enough related that all have the same scientific name of *Sus scrofa*. Numerous common names such as razorbacks, piney woods roosters, wild boar, feral hog, feral pig, wild swine, and others all refer to the same species, although there may be differences in appearance.

Three million feral pigs reside in the state of Texas, and some of these are crossed with European wild boars that were imported by sportsmen for private hunting reserves and escaped. All the different varieties do massive amounts of damage to many different crops. Hunting season on feral hogs is wide open in the state, no license required, and no bag limit. Kill all you want, the more the better. Nevertheless, sportsmen haven't even put a dent in the population.

Now here is the conundrum: There are lots of feral hogs on Cocos Island, they used to be hogs on Caño Island, and North America has an enormous problem with them. Why aren't there any on mainland Costa Rica?

You may say that predators like jaguars eat them. Yet in 2018 I had the good fortune to visit Pantanal National Park in Brazil, home to the highest density of jaguars of any place in the world, and we saw feral hogs in the park. I don't have the answer, but I love pondering Mother Nature's brainteasers. My guess is that the few feral pigs that may have been introduced onto mainland Costa Rica were killed by people and used to make tamales.

36. Hunting and Poaching

Human Beings: Top Predator

At about 8 years of age I started going with my dad on Sunday mornings to his skeet club in Greeley, Colorado. I was too young to handle a gun, but loved watching him and his friends blow flying discs called "clay pigeons" out of the air with their shotguns. Even though I wasn't shooting, my dad hammered firearm safety into my head on those Sunday morning outings.

My 12th birthday present was a 28-gauge shotgun. My first victims were doves, but I quickly advanced to ducks and geese. My dad loved hunting Canadian geese and took me with him two or three times each year. I bagged my first goose at age 14. By the time we immigrated to Costa Rica in 1970 three large game trophies adorned the walls of our house, one set of deer antlers and two bearskin rugs. All of these were killed in Colorado and Wyoming, and all were legal, licensed kills. My father and grandfather both taught me to abide by hunting laws and regulations, and, above all, to respect private property. "Always ask permission to hunt from the property owner," they told me. "If he says 'no,' don't trespass."

Costa Rica

In Costa Rica in the early 1970s I once went hunting, unsuccessfully, for paca with some friends. Within a few short years I had become enchanted with the rainforest and learned to look at nature and wild animals in a new manner. Not only did I give up hunting myself, but did everything in my power to stop people from hunting on Hacienda Barú. By the early 1980s I was starting to think of myself as an environmentalist. From that time forward Mother Nature and all of her natural checks and balances have never ceased to amaze me.

Prior to 2012 it was legal to hunt deer, doves, peccary, and pacas during limited and well-defined seasons. In that year a modification to the Costa Rican Wildlife Law #7317 made all hunting and poaching in Costa Rica illegal. It also prohibited the capture and holding in captivity of birds and other animals. The only activity that has ever resembled trophy hunting in Costa Rica should more accurately be called "trophy poaching." Unscrupulous hunting guides have charged foreign hunters upwards of $5,000 to kill a jaguar in Corcovado National Park. The result has been plunging jaguar numbers there. To my knowledge none of these poachers were ever caught and punished. Prior to the passage of the new law, the fines would have been insignificant. Now the maximum penalty for poaching is four months in prison and a $4,000 fine, still much too low, in my opinion, for killing a jaguar.

I believe that the law prohibiting all kinds of hunting and poaching is ideal for Costa Rica. Populations of many of the species once hunted and poached are threatened or endangered and need to be protected. Costa Rica has set an example for the world by protecting 25 percent of its land in parks and nature reserves and by doubling its forest cover in the last 30 years. Today 52 percent of the national territory is covered with natural forests. The country is known for its leading role in ecotourism. Tiny Costa Rica has as many species of birds as all the United States and Canada combined, and about 5 percent of the world's biodiversity.

The new law is another example of the country's commitment to conserve nature. It is overwhelmingly popular with the public. An environmental group acquired 177,000 signatures on a petition to the legislature requesting the passage of the law. It passed quickly and unanimously. On a practical level the law provides the legal framework to deal with poachers on the rare occasions when they are caught. What is lacking is funding for the enforcement of the law. In the meantime, poaching is still rampant in much of the country. It is up to the individual land manager or owner to defend their own property, because little help can be expected from the underfunded wildlife department. Hopefully the government will look for a way to fund the enforcement of this popular law.

Colorado

Though I believe the law prohibiting all hunting is great for Costa Rica, it is not right for everywhere. An excellent example is the state of Colorado, where sound wildlife and fishery management has been in place for many years. Wild animals are seen as a renewable resource that can be harvested sustainably. Population numbers of large ungulates, like deer and elk, are estimated. Their natural food supply is also estimated and hunting licenses are issued accordingly. Colorado is home to the largest elk herd in the world, over 280,000 animals. This number is considered optimum for the amount of food available to them in their natural habitat. About 208,000 elk hunting licenses were sold in 2021. This sounds like way too many, but wildlife managers know from experience that less than one out of five hunters who buy licenses will actually kill an elk. This reduction in population will be offset by new births, and next fall the population will be about the same.

In Colorado it is legal to hunt deer, elk, pronghorn, bighorn sheep, mountain goats, moose, pumas, black bears, ducks, geese, turkeys, and pheasants. A license is required to hunt and there is a well-defined season and bag limit. Every year about 575,000 hunting licenses are sold in Colorado, 86,000 to nonresidents. Nonresident licenses are much more expensive. For example, an elk license for a resident costs $49 and a nonresident has to pay $600. In order to purchase any license, you must show proof that you have taken a special course in hunter education and safety.

Each year over $80 million worth of hunting licenses are sold in Colorado and another $28 million in park entrance and camping fees. All this revenue goes to the Colorado Parks and Wildlife Department to help cover their operating expenses. Among other things, this revenue makes it possible for the department to vigorously enforce the law.

In addition to license fees, if you figure in the value of all the goods and services sold to hunters and fishermen by Colorado businesses, the value of the activity comes out to about $2.8 billion annually and 27,000 jobs created. Wildlife watching (ecotourism) brings in another $2.2 billion and generates 19,500 jobs. For more facts and figures, search the Colorado Parks and Wildlife website.

Needless to say, hunting and fishing is very important for the economy of the State of Colorado. I would imagine that any Colorado legislator who proposed a law similar to Costa Rica's prohibiting all hunting would probably be strung up from the tallest tree on the capitol building's front lawn.

Africa

The killing of the famed Cecil the Lion in Zimbabwe focused lots of attention on trophy hunting in Africa. Some airlines have gone so far as to refuse to transport hunting trophies from animals killed by foreign hunters. The public outrage over the horrible killing of Cecil has been overwhelming, and rightly so. I won't go into the gory details. I'm sure most of you have already heard them.

Nevertheless, responsible trophy hunting (hunting for sport, not food) is not necessarily bad for animal populations and local communities. We have seen the example of elk hunting in Colorado. There are many countries in Africa that permit trophy hunting in one form or another. Some countries have poorly managed programs that produce little income for local communities and animal numbers constantly decline.

In addition to legal trophy hunting, we have all heard of the toll that poaching is taking of African wildlife, especially elephant and rhinoceros. Several studies searching for solutions to the poaching problem have made some important discoveries. Most poachers are from local communities. They are poor and have families to support. They receive relatively little for the tusks and horns they poach compared to those above them in the chain of traffickers. Nevertheless, it is a source of income that allows them and their families to live relatively well compared to their neighbors. In addition to poaching, people often kill wild animals which damage their crops or kill their livestock. Rather than seeing wildlife as an asset, they see it as a detriment.

On the other hand, countries like Namibia have programs that produce positive results. In the mid-1990s Namibia initiated a system that guarantees local communities a say in the management of land and wildlife and a share in the monetary benefits derived from both ecotourism and trophy hunting. The program established local organizations called Community-Based Natural Resource Management (CBNRM) conservancies. According to Jason G. Goldman, writing in *Conservation Magazine*, "The CBNRM finally allowed local communities to benefit directly, both socially and economically, from the wildlife with whom they coexisted."

The CBNRMs receive a percentage of the income from lodges and trophy hunting outfitters. The percentage is negotiated between the private business and the CBNRM, so there is some variation from one area to another. The amount paid by lodge owners averages about 10 percent of the lodge's income, whereas hunting outfitters must pay between 30 percent and 75 percent of their income to local conservancies. The percentage varies depending on the species being hunted. Additionally, lodges and outfitters are required to hire a certain number of community members. In all cases the trophy hunting brought in more money, but the ecotourism produced more jobs. The funds acquired by the CBNRM are used to pay expenses which include salaries for wildlife guards, park managers, administrators, vehicle maintenance, fuel, etc. The profit goes to community projects.

Robin Naidoo's study determined that 74 percent of the CBNRMs studied were operating in the black and 26 percent in the red. He did a simulation to find out what would happen if the revenue from either hunting or ecotourism were eliminated. The result was that if ecotourism were eliminated, the percentage of profitable CBNRMs would drop to 59 percent, whereas if trophy hunting were eliminated only 16 percent would be able to pay their expenses. Without the CBNRMs a land area the size of Costa Rica with all its wild animals would go without protection.

These results are very encouraging. Namibia's wildlife was in decline prior to the initiation of the program in 1997. Since then, poaching has diminished, the killing of pest animals by local farmers has diminished, and wildlife populations are stable or increasing.

Killing Animals

I speak to lots of groups of tourists from North America, and always mention Costa Rica's no-hunting law. Most visitors are very favorable to the law, but there are always a few in each group who tell stories of how wild animals have caused problems for them and their neighbors. This includes everything from deer invading and damaging gardens in residential areas to coyotes killing sheep. They always ask me if I would be against killing wild animals in such cases. I wouldn't do it myself, I tell them. I no longer have a taste for that sort of thing. Costa Rica and the rainforest have changed me. Now I cringe every time one of our cats kills a gecko. But I certainly respect your right to hunt, if you don't infringe on the law and don't trespass on other people's property.

Photo: Pacas are relentlessly pursued by poachers.

37. The Shell Game

Get in Line and Stay in Place

"Hey Eduardo! Look. Somebody cut this coconut husk open, and it is full of all these little crabs. Boy, do they stink."

"It's not the crabs that stink," he laughed. "It's the rotten coconut. They love it. When I need fish bait, I just look for a coconut husk that's been cut open. If there's still any rotting coconut meat left, it's sure to be full of colonchos." He stuck his hand through a hole in the coconut and pulled out one of the colonchos, which I later learned are called hermit crabs in English. Eduardo held tight to the shell, a little smaller than a ping pong ball, grabbed the crab's head and pincher, and pulled steadily until the little invertebrate came free of its protective home, which Eduardo dropped on the ground. The head looked like a crab and the back part more like a fat worm. "Fish love them," he said with a smile.

Reaching into his bag, Eduardo pulled out a hand reel with some light fishing line and a small hook which he stuck through the crab and tossed it into the estuary. It didn't take long before he had a nibble. After several tries, he hooked a fish that he called a machaca. In half an hour he had four. They weren't much bigger than his hand, but he took them home anyway. "My wife knows how to fry them up real tasty."

I once heard a story about a delightful little beach (we will call it Pretty Beach) lined with coconut palms and Indian almond trees. Away from the hot sand, in the shade of the trees were lots and lots of hermit crabs. One day a stranger came to Pretty Beach and offered the people who lived there two hundred colones for each hermit crab they brought him. The adults had other, better paying, work to do elsewhere, but the kids ran right out and started gathering up coloncho for the stranger.

Good to his word he paid 200 colones for each crab. It was easy to find the crabs, and it only took five of them to make a thousand colones, enough for a soda or a bag of chips. The buyer left with several hundred crabs and said he would be back in a week. When asked what he was going to do with so many hermit crabs, he smiled and said he was going to sell them as pets. "Who would buy a coloncho for a pet," they asked incredulously, and he asserted that in North America lots of people would pay good money for them, give them names, and paint fancy designs on their shells.

The kids from Pretty Beach all thought that was the craziest thing they'd ever heard, but they kept on bringing him crabs, and he kept on paying for them. Before long, it became harder and harder to find hermit crabs until one day the stranger showed up and the kids had less than a dozen for sale. That was the last time they saw him.

It wasn't long before the beach started to smell like rotten coconuts and dead fish. Before the arrival of the stranger, the hermit crabs had eaten almost any kind of organic matter and kept the beach clean, but now they were all gone, and the bad smell lingered. There was no more fish bait either. The people thought that before long the ocean would bring more hermit crabs, but that didn't happen either. Pretty Beach was never the same ever again.

Why didn't newly hatched hermit crabs colonize Pretty Beach? Because there weren't any shells for them. The crabs hatch in the ocean and go through several growth stages, but once the larvae become mature hermit crabs, they come out of the sea onto the beach and look for a shell in which to live, usually one that belonged to another hermit crab. The crabs that don't find shells fall prey to one of many hungry predators. It is important that the shell be the right size, and as the crab grows it must acquire new and bigger shells. Let's say a hermit crab is feeling a little tight in its shell and decides to get a new one. It starts looking around for a bigger one. Finally, it finds one and tries it on for size, but it happens to be a little bit too big.

You would imagine that the crab would keep looking, but that's not what it does. Instead, it waits beside the bigger shell until a second crab comes along looking for a new shell. Number two tries on the empty shell, but as luck would have it, the new shell is a tight fit and won't work for it either. So, both crabs wait beside the empty shell until a third one comes along.

This goes on, and the crabs form a line according to size until finally one comes along that fits the empty shell perfectly. It moves in and abandons its old shell which the others then start trying on for size. With lots of good fortune everyone ends up with just the right shell, and there is one empty shell left for some other lucky crab. But sometimes the struggle for just the right shell gets violent and can even end up with the death of one or more crabs.

The shells are the key to the life of a hermit crab. Please don't take them away from the beach, and even more important, please, please, please don't ever buy a pet hermit crab.

Unfortunately, Pretty Beach is doomed to be lined with stinky coconuts for ever and ever. All the shells have gone north.

38. You've Got To Be Tough To Be an Agouti

What a Cool Cave, Let's Show Mom

When you're walking through the rainforest and surprise an animal that goes bouncing through the undergrowth emitting panicked, bark-like grunts with every bounce, you've almost certainly startled a Central American agouti (*Dasysyprocta punctata*). After 20 years of living here I finally saw one standing still. Possibly because once hunting was under control on Hacienda Barú National Wildlife Refuge, their fear of humans diminished considerably.

Some say that agoutis look like rabbits with short ears. For me their shape, habits and mannerisms are more like those of large, tailless squirrels. Like squirrels during times of plenty, the agoutis bury seeds. Though rainforests produce an abundance of seeds and fruits, there is always a time of scarcity. These hoarded seeds can mean the difference between survival and starvation.

Agoutis have figured out that white-fronted capuchin monkeys are very wasteful and dribble lots of goodies on the ground, so they often follow along underneath a foraging troop of monkeys and salvage the edible refuse.

Baby agoutis are the only newborn mammals I know of that actually select their own den site, separate from that of their mother. This happens the day after birth. Their den is so small that the mother agouti can't get inside and has to call them to come outside and nurse. At about three weeks the newborns start following mom around and learn what they need to know to face the world alone. Once the mother agouti determines they are ready to fend for themselves, she chases them away. Mother Nature applies the law of the jungle to these youngsters unmercifully; only the fittest and luckiest survive, a mere 30 percent. Their two main challenges are avoiding starvation and predation by coatis, ocelots, pumas and others. You've got to be tough to be an agouti.

In 2015 one female agouti got brave enough to leave the protection of the forest at dusk, scamper across a driveway to Hacienda Barú Lodge, grab a fallen mango, and hurry back to the jungle. Each day she made her jaunt a little earlier, without incident. She soon lost all inhibitions and could be seen in broad daylight sitting under the tree pigging-out on mangos.

Her offspring learned from a young age that there was nothing to fear from humans and got in the habit of wandering around in the gardens. Feeding them is prohibited at the lodge, but they have learned where all the fruit trees are located and check them out regularly. This has been going on for about a dozen generations.

If for some reason these semi-tame agoutis were forced to return to the forest and live like wild agoutis, they wouldn't stand a chance of survival.

39. Going, Going...Gone

And When They're Gone, They Don't Come Back

I love Costa Rica and its wildlife, and sometimes I envy the pioneers who lived here in the early 1900s and were able to fully appreciate some of the wildlife that are no longer with us today. I am thinking of two species in particular, the giant anteater and the harpy eagle, truly magnificent species that once lived in Costa Rica. Today the former is almost certainly gone forever, and there still may be a few of the latter in several remote locations in the country.

Giant Anteater (*Myrmecophaga tridactyla*)

This large mammal is similar in many respects to the anteater many of us know today as the tamandua (*Tamandua mexicana*): both have long curved noses and long curved claws that tuck completely under like a fist and obliges them to walk on their knuckles. Both eat ants and termites. That's where the resemblance ends. A large tamandua can weigh up to 5 kilos and is about 60 cm long, and the giant anteater, sometimes called the "ant bear," weighs eight times as much, around 40 kilos, and is more than double the length. In Costa Rica they are called *oso caballo*, which literally translates to "bear horse," presumably because of its large size and big, round, furry appearance.

I had the good fortune to see two giant anteaters on my once-in-a-lifetime trip to the Pantanal in Brazil. Both were at night, and neither would let me get close enough to use a flash to take a photo. Our guide told us that they eat mostly ants, first digging an opening into the ant hill, inserting the long narrow snout as far as possible, and then inserting a sticky, snake-like tongue, which is as long as a man's arm, deep into the nest to retrieve the ants. He said that the anteater can flick its tongue in and out at the awesome rate of about twice per second. After a very short time it moves on to another nest. This may be because the soldier ants guarding the nest from intruders quickly come out in force. With their long powerful pinchers, they may be able to penetrate the thick hairy hide that protects the animal from ordinary ants.

The last confirmed sighting of an ant bear in Costa Rica was in Guanacaste in 1984 and there were several sightings in the Peninsula de Osa during the 1960s and 70s, but these beautiful mammals are almost certainly all gone now. My good friend Patrick O'Connell, better known as the "Gold Walker," walked over every corner of the Osa Peninsula for 26 years, beginning in 1966, buying gold. He saw a tremendous amount of wildlife but never saw a giant anteater nor heard of a sighting from a reliable observer.

Some sources say that they were hunted for their fur and meat. I imagine that they also kill hunters' dogs and hunters retaliate by killing ant bears. The much smaller tamandua is capable of meting out very nasty gashes with its long powerful claws when attacked by dogs. I can only imagine what would happen to a dog if a giant anteater got a hold of it. There are several recorded cases of giant anteaters killing humans that got too close and tried to mistreat the animal. In the Pantanal we learned that jaguars won't mess with a giant anteater.

Harpy Eagle (*Harpia harpyja*)

It is possible a few of these majestic birds of prey still can be found in remote areas of Costa Rica. If so, it is probable that wildlife authorities are not divulging the information to the general public. A park ranger in Corcovado National Park told me many years ago that they had been monitoring a nest with one healthy chick that was getting close to fledging. One day the chick disappeared. Later they heard rumors that it had been stolen by a poacher and sold to a foreigner for a great deal of money. This type of poaching may be the greatest threat to the harpy eagle. How much would an unscrupulous collector or dealer be willing to pay for a healthy chick of the largest and most powerful raptor in the world? I can only imagine.

About 20 years ago a young lone male was reported in Corcovado. It flew in one morning and landed in a tree near the Sirena Biological Station. A biologist friend, one of about a dozen witnesses including park guests and rangers, told me that when the huge bird of prey landed in a large tree at the forest edge, a group of spider monkeys foraging in the crown of a tree within sight of the eagle literally plunged into the lower branches and cowered there until it flew away. It had been many years and several generations since any monkey in the park had seen a harpy eagle, yet the fear was so embedded that they immediately perceived the threat and reacted accordingly. Harpy eagles' primary prey are monkeys and sloths which they snatch out of the crowns of rainforest trees.

It is known that there are still a few in Panama, Venezuela and Surinam. Hopefully we still have a few in Costa Rica.

Jabiru Stork (*Jabiru mycteria*)

Any discussion of extinction in Costa Rica should include the jabiru stork. There are still a few left in the wetlands of the Tempisque Basin, but the population is so small that it is probably seriously inbred and will not likely survive. Ornithologists are monitoring them closely. Several years ago, there were still 23 individuals including several nesting pairs. Though wildlife personnel were monitoring the nests closely, one adult disappeared from the nest where it was incubating an egg from one day to the next.

Devastated by the news, the wildlife inspectors launched an investigation. Several days later they found feathers near a farm worker's home. The man admitted that he had killed the grandiose bird and he and his family had eaten it. I'm sure he has no idea of the true value of that meal.

40. A Flashlight and a Roll of Toilet Paper

If You Have These, You Will Survive

Costa Rica was still a third-world country when I first came here over 50 years ago, and part of me wishes it still was. Some of my most treasured memories are from that period in the early 1970s when I worked on a large ranch on the Caribbean side of the country. It was only accessible by a narrow-gauge railroad called the Northern Railway. I hadn't been there long when I met a great guy, an Englishman by the name of Johnny James who came to Costa Rica in the 1930s, before the concept of the Third World had even come into being. I once accompanied him on a four-day adventure, during which I learned much of what I know about the Caribbean side of the country.

It all began in a place called El Cairo in a restaurant owned by Juan Quiros Ching. No sooner had we finished our coffee and empanadas than the owner of the burro carro, our mode of travel for the first leg of the journey, walked in and shouted to Johnny that he was ready to go. The burro carro was a small platform with four railroad wheels. It was drawn by a mule. During the hour it took to reach the creek at the end of the spur, Johnny talked non-stop.

"Years ago," he began, "the immigration of Chinese to Costa Rica was prohibited by law. But what can I say? You know how it works. Laws are for common people like you and me, not for the wealthy, especially not back then. At the beginning of the century lots of Chinese were brought here to work as indentured servants in the households of well-to-do Costa Rican families. After a certain number of years of unpaid service, their debt was fulfilled, and their masters had to free them. Most took the last names of those same former masters and eventually acquired Costa Rican citizenship. Just like Juan Quiros," Johnny continued. "He and his wife were betrothed as children and came to Costa Rica from China at the age of ten or eleven. To put it bluntly, their parents sold them to a rich Tico. They had to work for fifteen years to earn their freedom. Then they moved out here to the jungles of the Caribbean, along the railroad, and proceeded to make their fortune. They worked their butts off, but today they've got the business, a farm, and a pile of money."

At the end of the spur, the man who brought us unhitched the mule, took it around and hitched it to the other end of the burro carro, and returned to El Cairo. A small boat with a tiny outboard motor was waiting. "We need to get going," urged the boatman. "It's getting late."

We had the river current in our favor and made it to Parismina just before nightfall. Dinner was delicious; steamed vegetables, rice, and chicken, prepared by the owner of a pulperia, a friend of Johnny's and fellow countryman of Juan Quiros. We slept in a couple of upstairs rooms. The next day we puttered on in our little boat through the canals to Tortuguero where a talkative lady, hungry for news, fixed us a tasty lunch of white-lipped peccary in tomato sauce with rice and beans which she served at a table on her front porch. From there we proceeded on to Limón where we boarded the southbound train.

During the train ride Johnny explained that the Atlantic railroad was owned by the Northern Railway Company, which was founded by a British industrialist named Minor Keith, the man generally credited with accomplishing the gargantuan task of constructing Costa Rica's first railroad. The project was begun in the early 1880s and completed in 1891. Partially financed by United Fruit Company, the railroad made banana production feasible in the Atlantic zone.

Keith soon found that Costa Rican workers had little resistance to malaria and yellow fever and would not be able to provide the labor for building the railway, so he brought in laborers from Jamaica. The Jamaicans were good workers and had a long history of resistance to tropical diseases. Had it not been for them, the Atlantic railway might never have been completed. Nevertheless, they were given little in the way of appreciation for their contribution to Costa Rica's development. Even Jamaicans born in Costa Rica were, for years, considered to be Jamaican citizens and were not afforded the same rights as Costa Ricans. At first, they were allowed to travel no farther west than Peralta, a small town between Siquirres and Turrialba. Later they were allowed to go as far as Turrialba, then Cartago, and finally to San Jose and the rest of the country. Eventually the Jamaicans were given citizenship and all the rights that came with it.

We stepped off the train at a place called Penhurst. Before we could continue to Cahuita, our destination for the night, we needed to cross the Estrella River. This was accomplished in a small boat called a cayuca, which was propelled by a boatman with a long pole. The bus was broken down, so we were met by a truck instead. I remember the driver brushing his teeth in the muddy river water.

A light rain started falling as we climbed into the back of the open-bed truck. I noticed that the tico truck driver and his two black companions were speaking a language I didn't understand. "It's a local dialect," Johnny explained. "The older Jamaicans speak only English and the younger ones English and Spanish, but everybody speaks *patois*, except us outsiders." About the time we arrived in Cahuita, the rain stopped. After a satisfying dinner of rice and beans and fish cooked in coconut oil by a delightful Jamaican man who was determined to sell me a lot on the beach, we slept in the only hotel in town.

The bus was working the next day and took us on a morning excursion through the countryside. It broke down again just outside of Puerto Viejo. We hiked into the village and found a guy with a Toyota pickup who took us back to Cahuita where we had a relaxing, uneventful afternoon. The next morning, we returned to Penhurst the same way we came. The rest of the trip home was by rail with one change of trains in Limón and another in Siquirres.

My four-day adventure with Johnny James ended in El Cairo at dusk. During dinner at Juan Quiros Ching's restaurant, Johnny offered a bit of wisdom. "You're a good listener," he remarked casually. "Now you know a bit more about this part of the country. I have a feeling that you're gonna be in Costa Rica for a long time, maybe even the rest of your life, and you're gonna have plenty of adventures. Just a bit of wisdom from an old-timer; wherever you go, take a flashlight and a roll of toilet paper, and you will survive."

ACKNOWLEDGMENTS

This collection of stories previously appeared in two English-language magazines in Costa Rica. I am grateful to David Bolger, publisher of *Quepolandia,* for his encouragement and promotion of my articles, and to his graphic designer Paul Rees, who worked his special magic on the photos I sent with the articles. The late Dagmar Reinhard, publisher of *Ballena Tales,* and Carlos Leon, editor and designer, were both helpful in choosing my subject matter and allowing me to exceed the word limit on more than a few occasions.

Some chapters include the stories of exceptional people: Patrick J. O'Connell, better known as The Goldwalker; Alejandro Gutierrez who shared the story of surviving a devastating flood; the late Thomas J. Brower, the first gringo in Dominical, whose story was shared by his children and grandchildren; the late Dr. Otto v. Helversen who gave me a great appreciation for bats; Randy Burns who told of his ordeal being bitten by a pit viper; and Ana Maria Torres, MsC, who gave me fresh insight into scarlet macaws and their return to this area. I greatly appreciate the contributions of each of these people whose absorbing stories add so much to this book.

Many thanks to my brother Rex who helped select which stories to include and kindly offered his proofreading skills once we had a final draft. It would take many pages to list everything that LaVonne Ewing did to make this book possible. Her editing and organizational skills are exceeded only by her skills as a designer and publisher. I am forever grateful for her talent and hard work.

Since I retired, I haven't had an English-speaking editor to help me shore up my articles, so I have been left to my own devices. Once I have what I think is a final draft I read it out loud to my wife Diane, only discover that it is far from final. Her suggestions make the articles flow in ways they otherwise wouldn't, and her contribution and support are greatly appreciated.

Thank you to Jan Betts for another wonderful cover illustration; I've enjoyed our delightful collaboration over the years.

I am thankful for the following talented photographers and the photos they have graciously shared: Rigoberto Pereira Rocha for the photo of the monkey eating an iguana and one of a jaguarundi, Sabine Bernert and Christal Stefani for extraordinary monkey photos, and Riccardo Oggioni for the sloth and the butterfly photos. Thanks to Dr. Mike Mooring from Quetzal Ecological Research Center for several photos from his many trail cameras located in parts of Costa Rica where tapirs, bush dogs, red brocket deer and other rare species are found.

Other photos were provided by Diane Ewing, Hacienda Barú's trail cameras, and my own small Cannon point-and-shoot camera. A few hard-to-find images (i.e. the satellite image of a hurricane) were sourced from stock agencies.

RECOMMENDED READING

Brower, David. *Let the Mountains Talk, Let the Rivers Run.* Gabriola Island, BC, Canada: New Society Publishers, 2000.

Coates, Anthony, ed. *Central America, A Natural and Cultural History.* New Haven: Yale University Press, 1997.

Edwards, Hugh. *Crocodile Attack.* New York: Avon Books, 1989.

Food Production and Population Growth. 2 hours, 40 min. With Daniel Quinn and Alan Thornhill, Ph.D. New Tribal Ventures, Inc., 1998. Videocassettes (2).

Giono, Jean. *The Man Who Planted Trees.* White River Junction, Vermont: Chelsea Green Publishing, 1985.

Hoyt, Erich. *The Earth Dwellers.* New York: Simon & Schuster, 1996.

Jansen, Daniel H., ed. *Costa Rican Natural History.* Chicago: University of Chicago Press, 1983.

Koonin, Steven E. *Unsettled: What Climate Science Tells Us, What It Doesn't, and Why it Matters.* Dallas, Texas: BenBella Books, 2021.

Leaky, Richard and Roger Lewin. *The Sixth Extinction.* New York: Doubleday, 1995.

Leenders, Twan. *A Guide to Amphibians and Reptiles of Costa Rica.* Miami, Florida: Zona Tropical Publishing, 2001.

Quinn, Daniel. *Beyond Civilization.* New York: Harmony Books, 1999.

_____. *Ishmael.* New York: Bantam Books, 1992.

_____. *My Ishmael.* New York: Bantam Books, 1997.

_____. *The Story of B.* New York: Bantam Books, 1996.

Shellenberger, Michael. *Apocalypse Never: Why Environmental Alarmism Hurts Us All.* New York: Harper, 2020.

Terbrough, John. *Requiem for Nature.* Washington, D.C.: Island Press / Shearwater Books, 1999.

Wilson, E. O. *The Future of Life.* New York and Toronto: Random House, 2002.

_____. *The Diversity of Life.* Cambridge, Massachusetts: Harvard University Press, 1992.

MEET THE AUTHOR

Jack Ewing's love affair with the rainforest began in 1970 when, in search of new opportunities for plying his Bachelor of Science degree and his skills as a cattle rancher, he left his native Colorado and moved his wife, Diane, and their young family to the jungles of Costa Rica. His ever-growing fascination with the rainforest, however, soon prompted his transformation into environmentalist and naturalist.

A natural-born storyteller, Jack's articles about life in the rainforest appear regularly in Costa Rican publications, and he often speaks to environmental, student and ecological traveler groups. He is currently president of the local environmental organization ASANA. His expertise on biological corridor projects is much sought after.

"What we must do to save the rainforest," says Jack, "is connect the parks, refuges and reserves with biological corridors and then teach the people how to make a living from these natural areas without damaging or destroying them. If we want conservation to work, we have to make it profitable."

Jack and Diane live on internationally acclaimed Hacienda Barú National Wildlife Refuge where he is currently retired.

ALSO BY JACK EWING

Monkeys Are Made of Chocolate: Exotic and Unseen Costa Rica

Where Tapirs and Jaguars Once Roamed:
Ever-Evolving Costa Rica

Hacienda Barú National Wildlife Refuge is located on Costa Rica's Southern Pacific Coast—a region of distinct natural beauty, where forest covered mountains rise up from the dramatic Pacific coastline. A fantastic variety of habitats, from wetland and secondary forest in the lowlands to primary forest on the highland coastal ridge can be found on 330 hectares (815 acres). Seven kilometers (4½ miles) of trails and 3 kilometers (nearly 2 miles) of pristine beach are waiting to be explored. Our mission is to protect the wildlife habitats of Hacienda Barú, while educating our visitors about its biological wealth.

Hacienda Barú National Wildlife Refuge
HaciendaBaru.com

OTHER ORGANIZATIONS:

ASANA
Asociación de Amigos de la Naturaleza
(Association of Friends of Nature)
asanacr.org

www.ingramcontent.com/pod-product-compliance
Lightning Source LLC
LaVergne TN
LVHW091145080826
845145LV00008B/2262

* 9 7 8 1 9 3 6 5 5 5 6 7 3 *